HEALING

Crystals and Stones

Publications International, Ltd.

Written by Sheryl DeVore Bailey

Images from Shutterstock.com

Louis Weber, CEO
Publications International, Ltd.
8140 Lehigh Avenue
Morton Grove, IL 60053

ISBN: 978-1-64030-468-0

Manufactured in China.

8 7 6 5 4 3 2 1

TABLE OF CONTENTS

INTRODUCTION

For billions of years, nature has been creating an array of colorful sculptures, many of which originated deep within the Earth's core.

These sculptures come in a cornucopia of colors and shades. Their shapes vary from hexagons and cubes to irregular patterns. They're found alone, in clusters, and embedded in others of their like.

These sculptures are crystals—art created not by man, but by the forces of nature.

Since ancient times, crystals have been mined, admired, and used as adornments and products. They've also been used for enhancing meditation and healing.

This book describes 64 crystals and stones suggested to have healing properties. Each page contains photos of these crystallized minerals and includes information on where they are found, the lore and legends behind them, and what crystalogists consider to be their healing properties.

It may seem like science pitted against sorcery—attaching specific qualities to non-living objects—but when you learn about how crystals form and the properties they have, you begin to understand that mystique and reality could be compatible.

AN IMPORTANT NOTE Healing properties of crystals have not been validated by science, nor can anyone promise that using them will create the effect you desire. Those who need medical care—both for the body and the mind—should seek professional help. Statements made in this book about crystal healing properties have not been evaluated or endorsed by the U.S. Food and Drug Administration. Do your own research and talk to your healthcare

provider before attempting to use crystals for healing. You could consider crystal therapy as an adjunct to traditional care, but not as a cure-all. Use this book as an introduction to the world of crystals, their beauty, their stories, and their potential.

FORMATION

Earth is made of rocks, which contain various minerals. At least 3,000—and up to more than 4,000—different minerals occur naturally on earth and can crystallize. Within the Earth's core, a hot liquid mixture of minerals called magma moves through nooks and crannies and sometimes rises above the surface, where it is then called lava. As the lava cools and solidifies, the minerals crystallize, becoming stable structures with specific atomic properties. Crystals, though not considered alive, can grow given the right amount of space and time, and they can then be reshaped and reformed under extreme pressure.

Crystals can be formed by minerals in hot springs rising to earth's surface or vapor in the vent of volcanoes. Even frost forming on a windowpane is considered a crystal. The rarest and most valuable crystals are referred to as gemstones.

You can hunt for crystals for free on some public lands and for a fee on private lands. Check ahead of time for locations, times open for digging, and tools you might need.

Of course, you can purchase your own crystals from various stores or online. The key is to research the seller and talk to customers before purchasing. Some crystals can be imitations, so ask the seller specifically where they obtain their products. Ask friends and relatives who use crystals for advice, and read books about the healing properties of crystals, rather than relying solely on the Internet, which can contain a lot of hype about these crystals.

MINING AND HUNTING

Miners can find crystals in caves, near rivers, and in rocks. It's important to note that mining crystals needs to be done ethically and without destroying ecosystems. Many who use crystals do research to ensure they are getting ethically-mined products, keeping in mind that specimen mining—searching for a small crystal—is different from commercial mining, which can include heavy blasting. Discoveries of new mining locations for high-quality gemstones as well as common crystals are continually being made. Poaching does occur in the world of crystal mining, just as it does with hunting animals.

LOCATION AND IDENTIFICATION

Crystals have been found all over the world. From topaz and quartz found in Brazil to tourmaline in Afghanistan and Madagascar, different crystals can be mined in different regions of the world.

Crystals can be used in their rough form or as polished stones—but the words "crystal" and "stone" are often used interchangeably.

Crystal colors vary and can be used for identification, but other properties such as luster need to be examined as well. Additionally, many different structures are found among crystals, and different

diamond-shaped patterns and are said to energize. Aquamarine and apatite are hexagonal and thought to help with organization. Crystals are placed on a scale of 1–10 for hardness, called the Mohs scale, with talc or soapstone being one of the softest minerals, and diamond one of the hardest.

Some crystals can be identified by their inclusions—materials, typically minerals, trapped within the main body. Inclusions are often thought of as imperfections, but they can add beauty to the crystal.

structures boast different healing qualities. For example, garnet and fluorite have cubic patterns and are touted to release tension and inspire creativity. Ruby and tourmaline have

Crystals are believed to vibrate in a specific pattern when electricity passes through them. As such, crystals have been used to make watches, lasers, cell phones, and other products.

SEVEN CHAKRAS

Crystal healers believe each stone has its own vibration pattern. These vibrations can interact with different frequencies within the human body and affect what are known as chakras—the seven centers of energy.

According to crystal healers, when the chakras are aligned properly, your body and mind are said to be balanced. If one or more of your chakras is unbalanced, your health can suffer. Using specific crystals can restore the balance, they believe. The seven chakras have specific locations in the body with different colors associated with them. They start from a line above

your head and run to the base of your spine. Each is associated with different aspects of your mind, body, and spirituality.

You can get in touch with your chakras by closing your eyes, imagining each one individually for a minute, and thinking of the corresponding color as a light moving in that area. For example, concentrate on the brow chakra at the center of the forehead and imagine an indigo light moving in that space.

CROWN
Color: Violet
Location: Above the head
Function: Connection to the universe

BROW OR THIRD EYE
Color: Indigo
Location: Center of the forehead, above and between the eyebrows
Function: Intuition and spirituality

THROAT
Color: Light blue
Location: Throat
Function: Personal truth balanced with kindness

HEART
Color: Green
Location: Center of the chest
Function: Compassion for yourself and others; kindness

SOLAR PLEXUS
Color: Yellow
Location: Above the naval and below the diaphragm
Function: Wisdom, confidence, power

SACRAL
Color: Orange
Location: Just below the belly button
Function: Identity, pleasure, creativity

ROOT
Color: Red
Location: At the tailbone or between the thighs
Function: Survival and safety

CHOOSING, CLEANING, AND USING

This book reveals properties associated with a wide range of crystals, but you can also search for a crystal that speaks to you by visiting stores or going online. Choose what feels and looks best to you. Focus on the color, shape, and size first before researching a stone's healing properties. Hold the crystal in your hands to see how it feels. Do you experience any warmth, tingling, or other sensations?

Crystalogists use stones in a variety of ways, depending on the purpose. They recommend cleansing them and setting an intention before use. When cleansing your crystal, make an intention to clear unwanted energy from the stone, and imagine white light surrounding it.

One cleaning method involves soaking the stone in warm salt water; another requires holding it beneath running water—but make sure your stone is not too fragile. Water can damage some stones, so research your crystal before using water. Water, especially distilled, works well when cleaning quartz crystals, while selenite and turquoise stones may dissolve in water.

Some practitioners use a smudging technique to clean their crystals.

This ancient practice involves burning herbs and holding the crystal in the smoke, imagining negative energy leaving the stone. Others use drums, gongs, and what are known as singing bowls to cleanse the crystal with sound. You can also place crystals on a piece of selenite overnight to cleanse and re-energize them. Some practitioners also believe stones should be charged by sunlight after they are cleansed. They suggest

leaving them in the sun during the day, but note that certain crystals should not be kept in the sun. Crystals such as amethyst, fluorite, and topaz may fade when exposed to sunlight.

Store your crystal in a silk pouch or other container to protect it when not in use.

Crystalogists recommend dedicating new stones before using. Here's one technique: Clean your crystal, then hold it in your hand, close your eyes, and experience its energy. Create your own dedication mantra, for example: "I dedicate this crystal to benefit all living things." Say the mantra several times.

You might want to research the crystal's meaning and then set an intention before using it. Create a prayer you can recite while holding it in your hand or placing it on one of your chakras. Some healers recommend placing a crystal underneath your pillow at night. Others suggest you carry it around in your pocket. You can also use crystals as pendants for necklaces or in a ring.

Beginners may wish to choose a clear quartz crystal, then sit quietly with the stone next to them, breathing in its positive energy. Another good crystal for beginners is selenite, which you can move from the top of your head to your feet in hopes of feeling rejuvenated and cleansed. Use amethyst as a decoration in your home or as a meditation tool. Beginners may wish to try citrine as well, placing it in a windowsill to gather natural light and instill optimism.

MEDITATING WITH CRYSTALS

Some crystals provide a gateway to meditation. Palm stones—crystals that can be held in your hands—are often used for meditation. Sit in a quiet place in a comfortable position where you won't be disturbed, hold your crystal in your hands, and close your eyes, making sure to breathe deeply. Focus on the sense of the stone, feeling its energy. Continue to breathe in and out while focusing on the crystal—if your mind wanders, direct it back to the crystal. You can meditate silently for a few minutes or half an hour with the goal of maintaining awareness and letting all thoughts float away. Some stones touted to be the best for meditation include quartz, celestite, amethyst, lapis lazuli, labradorite, and selenite.

Stress and anxiety can be constant companions in the 21st century. Meditation with certain crystals has been said to reduce the stress and instill a sense of peace and calm. One stone that works well, according to practitioners, is desert rose selenite. They recommend holding the selenite in both hands and then sitting on the floor or walking slowly, whichever feels best. As you sit or walk, move the crystal in your hands to help you realize that worry and anxiety won't help you solve problems. You can also say a mantra such as, "I can see the light." You'll know when it's time to stop the meditation when you feel less anxious.

Two crystals, danburite and blue chalcedony, can be used to reduce anxiety and imbue a sense of well-be-

ing, according to crystalogists. They suggest holding one of each in your hands. Meditate on the feeling you get from the crystals, noticing similarities and differences. Then breathe deeply in and out several times, exhaling anxiety and inhaling lightness. Imagine a sense of calmness you have felt in the past and trust that you have no worries. End the meditation when the time feels right.

Other practitioners suggest placing lepidolite over the third eye chakra or holding a piece in each hand to reduce anxiety. Another suggested technique is holding black tourmaline in your hands and visualizing your anxiety and stress filtering through the stone. You can also set black tourmaline next to you while at work to ease stress, according to crystalogists. Other calming crystals include fluorite, aquamarine, shungite, and blue lace agate. Practitioners suggest searching for the crystal that best resonates with you.

Today we are bombarded with electromagnetic radiation emissions from devices like cell phones and computers. Though the jury is still out on if these rays are harmful to humans, some crystalogists believe certain stones protect against potential negative effects from electromagnetic radiation. Crystals said to counteract those effects include amazonite, hematite, sodalite, and black tourmaline. Placing these stones near a computer or other electronic device is recommended to reduce the rays. Before placing any crystals near electronic devices, make sure the stones are not highly magnetic so you won't harm your hard drive.

CRYSTAL GRIDS AND LAYOUTS

Many different layouts of crystals can be placed atop and around your body to achieve certain results. Additionally, grids can be created in various spaces in your home, at work, and outdoors.

To balance your chakras, chose one stone representing each of the seven chakras, then place them on the appropriate places of your body, one at a time, beginning with the root chakra. Stay with the crystals for about half an hour, and then remove them one by one, beginning with the crown chakra.

Practitioners suggest different crystal layouts to improve emotional and physical health. To enhance physical health, they suggest using one black, brown, or red crystal; two brown crystals; two green crystals; one yellow crystal; and one quartz crystal. Place the black or brown crystal below your feet, then place a brown stone next to each hip. Continue with a green stone on either side of the shoulders and a yellow crystal on the solar plexus chakra. Finally, put the quartz above the head. Lie on the floor with these stones for about half an hour and then take them off in the opposite order in which they were placed.

Crystalogists also recommend creating a sacred space inside or outside the home where you can resonate with stones of your choice and design a layout that feels safe and comforting to you. You can use this sacred space during meditation

or when seeking peace, solitude, and comfort.

Practitioners also suggest creating a crystal grid, a grouping of different stones that reflect your intentions and needs. You can purchase inexpensive cotton cloth and wooden grids upon which you place the crystals, or just create your own grid. Here's a basic way to create a crystal grid for your home. First, choose the location for the grid. Then, write your intention on a piece of paper and consider the grid design. A simple circle is useful when seeking protection and safety; a triangle helps those seeking expansion and growth; the Star of David is a symbol for the heart chakra and useful for balancing and decluttering.

Choose a center crystal and some tumbled stones said to work well with your specific goals. Also be sure to use your intuition and see which stones speak to you when choosing those that will go into your grid.

Finally, select a clear quartz stone with a pointed edge to activate the grid. Crystalogists call this stone a "quartz crystal point."

Some grid creators place the written intention in the center where the main stone will go. Then, they think of the intention as they place the stones they've chosen in a series of circles around the center, beginning with the outer edges and working toward the middle. The placement of the center crystal completes the grid. They then use the quartz point to draw invisible lines to connect all the crystals. Practitioners suggest leaving your crystal grid in its sacred space and sitting with it when the timing feels right.

Endless possibilities exist for crystal grids. The key is to choose stones based on your needs and to use your intuition to guide you to the right combination.

ABALONE SHELL

Chakras: Heart, Third Eye, Crown

NAMING, LOCATION, AND HISTORY

Abalone are sea creatures that live inside hardy shells made of calcium carbonate. The iridescent shells are typically found on the western coast of North America and in South Africa, coastal New Zealand, Australia, and Japan. In one cave in South Africa, archaeologists have discovered abalone shells deposited 100,000 years ago.

MYTHS AND LEGENDS

Though not a crystal, abalone shells have been used in ancient and modern times to heal and work with the chakras. Ancient people believed abalone shells blessed their owners with protection, emotional stability, and peace. Native Americans burned sage in abalone shells during their smudging ceremonies.

PHYSICAL PROPERTIES

The inside of abalone shells, with the iridescent shades of blue, green, pink, and other hues collectively called mother of pearl, seem to mimic the colors of the ocean.

USES IN HEALING

Crystal healers say holding an abalone shell will imbue a sense of tranquility, and it can help soothe and redirect the spirit for those dealing with emotional upheaval.

After a fight with a loved one or before you embark on a new adventure, such as a new job or vacation, you can place dried sage or loose leaf sage into an abalone shell. Light the sage and wave the smoke over your body to dispel any negative energy.

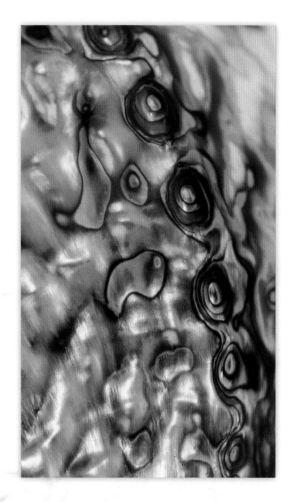

AGATE

Chakras: All

NAMING, LOCATION, AND HISTORY

The name "agate" is derived from the Achates River in Italy, where these stones have been discovered. Some records indicate agate was found in France as early as 16,000 BC, and agates found in Australia are said to be nearly 3 billion years old.

MYTHS AND LEGENDS

Ancient peoples often believed agate could turn a human invisible and protect man from evil, and some legends say agate can cure scorpion stings and snake bites and even stop thunder and lighting.

The ancient Chinese wore agate jewelry for good luck and to bring energy and purification to the body. During the Middle Ages, humans would tie agate to the horns of an oxen to ensure a good autumn harvest, and from the 1st through 19th centuries, Germans used agates to create cut and polished stones.

PHYSICAL PROPERTIES

Agates come in a variety of types and colors—from pink to blue to green to orange to purple to white.

True agates have some sort of banding. For example, they could have alternating stripes of quartz and chalcedony. One of the most beautifully banded agates is Laguna agate, a scarlet-colored crystal found in a remote area in Mexico.

Blue lace agate and Botswana agate are true agates because they have some banding. Tree agate is not, but it is included in the agate grouping among crystal healers.

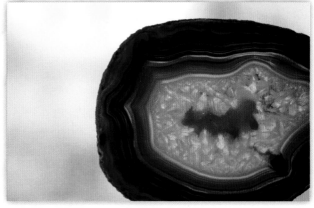

USES IN HEALING

In general, agates are said to strengthen and stabilize those who use them as well as balance their emotions and bring out their intellect. Agates have a lower intensity than other stones and vibrate on a slower frequency.

BLUE LACE AGATE

Chakra: Throat

NAMING, LOCATION, AND HISTORY

Found only in South Africa, Namibia, and a small location in Romania, blue lace agate is named for its lace-like appearance.

PHYSICAL PROPERTIES

Blue lace agate is an opaque crystal with bands of light blue and white, described as appearing like lace.

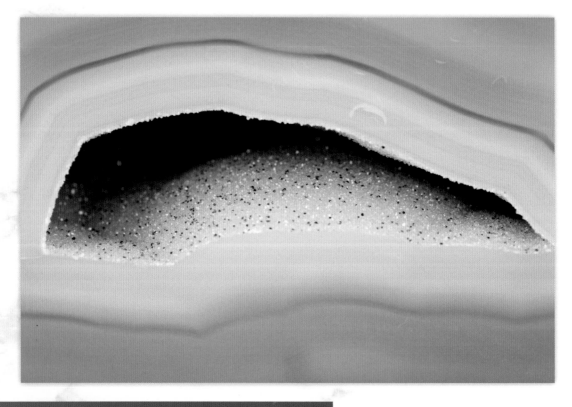

USES IN HEALING

Blue lace agate is said to get rid of self-doubt as well as provide confidence in achieving your goals, but this will take time since this crystal works slowly, according to crystal healers.

Blue lace is also said to be effective in helping forgive someone with whom you are angry. Holding the stone in your hands while thinking of that person can reveal the truth of the situation and release negative feelings, some healers say.

Blue lace agate is touted to help relieve sore throats, laryngitis, and other similar ailments, as well as arthritis and skin problems. Some have suggested it can calm hyperactivity in children.

Crystal healers also think that blue lace agate's association with the throat or the voice helps you speak the truth confidently without the fear of being judged. It's also said to help those who wish to improve their public speaking skills.

Overall, it's considered a stone of encouragement.

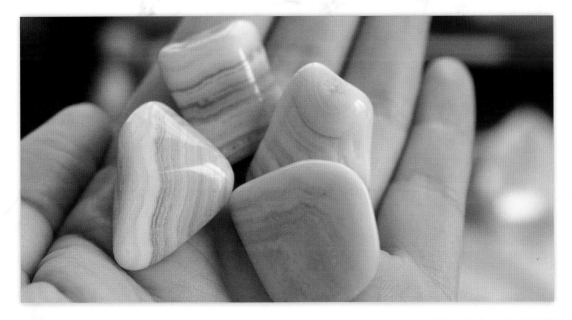

BOTSWANA AGATE

Chakra: Root

NAMING, LOCATION, AND HISTORY

Botswana agate stones were likely created more than 180 million years ago in Botswana, when lava flowed across the landscape and rocky layers. Africans have used Botswana agate in ceremonies to stimulate fertility and the production of healthy children.

PHYSICAL PROPERTIES

Botswana agate has bands of pink and gray, sometimes with brown or apricot hues as well. Some stones have "eye" formations within the crystal, which are said to bring good luck.

MYTHS AND LEGENDS

More than 3,000 rock paintings dating back as far as 1000 BC have been discovered near where Botswana agate has been found. One legend says a serpent lives at the bottom of a mile-deep whirlpool near this location. Unlucky humans who get too close can be grabbed by the serpent and taken into the undertow. If a person is wearing Botswana agate, he or she will be protected from the serpent.

USES IN HEALING

Considered a comfort stone, Botswana agate has been said to be useful for treating depression, anxiety, and panic attacks.

Botswana agate has been placed in vehicles as a protection against accidents, and some have claimed that it can ward off spiders. It may also reduce fevers when placed on the head.

Soaking Botswana agate in water for a few hours and then drinking the water is said to help those who want to quit smoking and give up other unhealthy habits.

AMAZONITE

Chakras: Heart, Throat

NAMING, LOCATION, AND HISTORY

Amazonite is found in granite rocks in the Jimensky Mountains of Russia, as well as in several other countries, including the United States. It has been discovered in Colorado near Pike's Peak.

Amazonite, sometimes called green feldspar, is likely named after the Amazon River in South America, where the blue-green stones have been discovered.

Jewelry containing amazonite and other crystals was discovered among King Tutankhamun's possessions in the early 1900s. King Tut lived and ruled during the late 1300s BC. Ancient Egyptians carved hieroglyphics on amazonite rock and wore jewelry made from amazonite, believing the crystal brought good luck and fertility.

MYTHS AND LEGENDS

According to legend, women of ancient civilizations in Brazil gifted amazonite to men who visited them.

In Babylonian mythology, amazonite was offered as a gift to Tiamat, the goddess of the sea, who was believed to be one of the first gods.

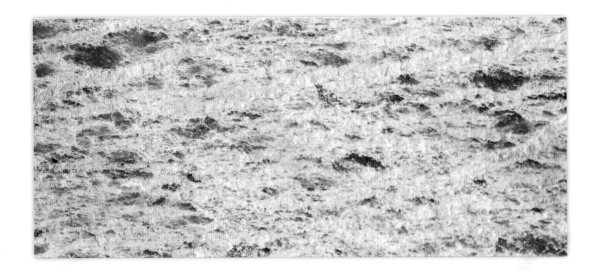

PHYSICAL PROPERTIES

Amazonite is a soft, blue-green stone. It has been suggested that its unique color comes from the presence of small quantities of lead and water in the stone.

You can purchase amazonite as a polished or unpolished stone and with or without white streaks or striations. The stone ranges from light green to bright green to blue-green.

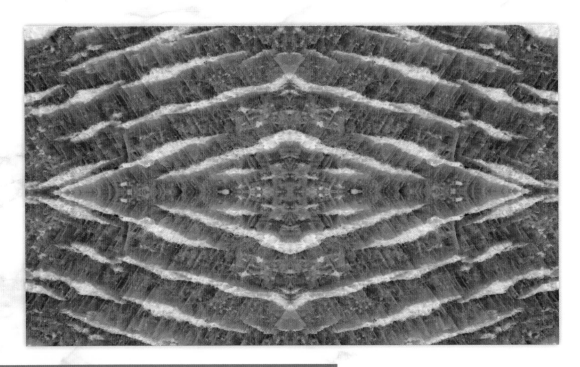

USES IN HEALING

Known as the stone of truth or the stone of courage, amazonite is said to focus on the heart chakra, which brings forth compassion and helps you know what you truly believe.

Celebrities such as Jennifer Aniston have worn amazonite rings, which are touted to be calming and promote healthy living.

Some crystal healers believe if you place amazonite between your heart and throat chakras and breathe deeply, its energy will calm your body and your mind, and it will release deep memories.

According to author Robert Simmons, holding amazonite tightly in your hand while speaking aloud your dreams can help make them come true. Other crystal healers suggest carrying an amazonite stone in your pocket to reduce worries.

When used with the throat chakra, amazonite is said to enable you to speak the truth with powerful words for the good of mankind.

Amazonite has also been said to lessen physical ailments—including gout and skin rashes—by rubbing the stone onto the affected area. Additionally, placing amazonite in a room with your computer and other electronic devices has been said to eliminate polluting rays. Combining it with black tourmaline or fluorite could strengthen this effect, according to some crystal therapists.

AMETHYST

Chakras: Third Eye, Crown

NAMING, LOCATION, AND HISTORY

Amethyst is found in South and Central America as well as the U.S., Canada, Russia, and Africa.

One of the largest amethyst deposits in North America was discovered in the late 1940s near Thunder Bay, Ontario, Canada. It's called the Amethyst Mine Panorama. Visitors can tour the mine and search for their own amethysts.

At the beginning of the 18th century, Queen Charlotte of England owned an amethyst bracelet considered to be priceless. Today, because more amethyst deposits have been found, its value has decreased. The rarest amethyst is called Deep Russian.

Amethyst has been discovered in Egyptian tombs, and bishops wore it on their rings during the Middle Ages, perhaps as a symbol of celibacy or dedication to Christ.

MYTHS AND LEGENDS

According to Greek mythology, when Dionysus, the god of intoxication, was insulted by a human, he decreed that tigers would eat the next person he met. That person was a princess named Amethyst. Amethyst was on her way to visit the shrine of the goddess Diana when she encountered Dionysus. One story says as the tigers attacked, Diana turned Amethyst into a beautiful clear crystal. Dionysus, sorry for what he had done, poured grape juice over the crystal, giving it a beautiful purple hue.

Another story says the princess hid in a crystal cave, which Diana then sealed to keep her safe. To make amends, Dionysus poured red wine on the cave, turning all the crystals purple, and Princess Amethyst was freed. In Greek, the word "amethyst" loosely means "not drunken."

PHYSICAL PROPERTIES

Amethyst's colors range from dark purple to reddish to violet to lavender to light purple. Mineralogists believe amethyst gets its colors from iron and manganese or aluminum.

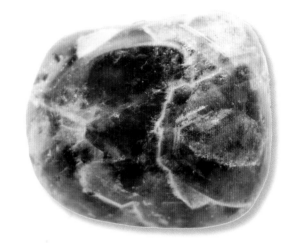

USES IN HEALING

Associated with the crown and third eye chakras, amethyst has been used to help people quit smoking and drinking. People also place amethyst in parts of their homes where they want to create an aura of protection and foster meditation or create harmony within the family. Amethyst is also used to cure headaches, keep emotions in check, and reduce stress. It's said to be able to rid the body of addiction as well as protect, purify, and bring one closer to a higher power.

AMETRINE

Chakras: Solar Plexus, Crown

NAMING, LOCATION, AND HISTORY

Ametrine, which belongs to the quartz family, consists of amethyst and citrine and is found in Brazil, Uruguay, and Bolivia. It's sometimes called bolivianite because of where it is mined.

MYTHS AND LEGENDS

A legend says a Spanish soldier who came with others to explore Bolivia met a peaceful tribe and fell in love with a princess member of the tribe. Her father bestowed a dowry of a mine with two-colored stones, but the husband-to-be was not impressed, preferring perhaps gold or emeralds.

When it was time to return to Spain, the conquistador wanted his wife to come with him—but her fellow tribesmen decided

to murder him so she wouldn't abandon her people. She warned her husband, and he fled, but the princess was injured by her own people in the process. Her husband returned, and she gave him a stone from her father's mine. She died in his arms

holding the stone of two colors— purple and yellow—to symbolize how she was torn between her love for her husband and her tribesmen.

PHYSICAL PROPERTIES

Usually transparent, ametrine's color contains both the purple of amethyst and the gold of citrine.

USES IN HEALING

Ametrine is touted to be helpful for those who want to lose weight and let go of bad habits as well as instill creativity and increase brain power. Crystal healers have also used ametrine to treat those who are depressed, anxious, or stressed.

APATITE

Chakras: Solar Plexus, Heart, Throat, Third Eye

NAMING, LOCATION, AND HISTORY

The word "apatite" comes from a Greek word meaning "deceive," likely because it exists in a variety of colors and can easily be misidentified.

The largest deposits of apatite are found in Russia, but some are also mined in Canada, Mexico, Brazil, Spain, and the United States. This crystal was likely created by the decomposition of sea creatures. It is naturally present in human tooth enamel.

PHYSICAL PROPERTIES

Apatite comes in many colors, including white, yellow, blue, green, pink, and multicolored. The different colors relate to different chakras.

USES IN HEALING

Blue apatite focuses on the third eye, and is said to improve psychic ability, helping you arrive at the moment of light when things become clear. It's also been said to help relieve headaches and vertigo as well as improve eyesight.

Green apatite relates to the heart, throat, and third eye chakras. This blue-green crystal is said to balance the mind and the head as well as engender gratitude. It's also been touted as useful for those with stress-related heart disease.

Golden apatite focuses on the solar plexus chakra. Crystal therapists sometimes recommend golden apatite for those who want to lose weight, determine what it is they want in life, or improve their digestive system. A blocked solar plexus chakra is said to keep people from accepting abundance, and golden apatite is used to unblock this chakra.

AQUAMARINE

Chakras: Crown, Third Eye

NAMING, LOCATION, AND HISTORY

Aquamarine's Latin name means "water of the sea," a fitting tribute to its clear, water-like, blue-green appearance. As such, it's linked to a cooling or soothing sensation.

Aquamarine can be found in Russia, Pakistan, South America, and the United States.

Egyptians considered aquamarine a symbol of everlasting youth. In Europe, in the late 1300s, people believed those who wore an aquamarine stone or pendant could counteract poisoning.

In the Middle Ages, aquamarine was one of the crystals most often used to create a ball for fortune telling. People back then also believed pairing an aquamarine with a diamond helped create a long, happy marriage. They considered aquamarine to be a perfect gift for a wedding anniversary, thinking that it would bring more love into the relationship.

One of the world's largest single pieces of aquamarine is housed at the National Museum of Natural History and weighs in at 900 carats. The gemstone, called the Dom Pedro aquamarine, was discovered in Brazil.

PHYSICAL PROPERTIES

Both aquamarine and emerald are different color variations of the mineral beryl.

MYTHS AND LEGENDS

Legend says mermaids bestowed aquamarine upon sailors to give them protection and luck. Another legend purports that Neptune, the Roman god of the sea, discovered aquamarine washed ashore and kept it close so it would not end up back in the water.

The Romans also believed a carving of a frog on aquamarine helped friends who had become enemies reunite. It was also used as a gift to wives by husbands upon the morning of their first day of marriage. A Roman naturalist and philosopher born circa AD 23 purported that aquamarine had healing properties for the eyes, and suggested washing the eyes with an aquamarine water to promote health.

Both Greeks and Romans considered aquamarine a symbol of safety when sailors were out at sea.

USES IN HEALING

Some modern crystal healers suggest creating a cleansing water using the raw form of the crystal to promote a youthful looking skin. The crystals need to be cleaned and then put into a glass bowl filled with filtered water. Setting the bowl of aquamarine crystals in the sun or full moon has been said to harness the power of beauty.

Aquamarine is also touted as a stone to heal mental angst, help people think more clearly, cast away a propensity for procrastination, and instill calm.

When focused on the throat chakra, aquamarine is said to soothe a sore throat, lessen laryngitis, and help alleviate allergy symptoms. Wearing an aquamarine pendant can return energy to the system, helping you avoid infections, according to some crystal healers. In addition, it's been said that holding aquamarine crystals close to you can help you overcome difficult challenges.

AVENTURINE

Chakras: Root, Heart, Third Eye

NAMING, LOCATION, AND HISTORY

Aventurine is found in Brazil, Russia, Canada, China, India, and Tibet. The word "aventura" means "chance," and as such, aventurine is said to bring good luck. Other stories, however, say the meaning of chance refers to the production of a glass in Italy in the 1700s that was dropped and then cooled to create a sparkling appearance. Aventurine is not that product, but it has a similar appearance, glistening from mica materials embedded within.

PHYSICAL PROPERTIES

Aventurine is a very hard stone. Because of this, humans have used it to make tools and other products for millions of years. It can be found in green, blue, and red.

USES IN HEALING

When aventurine is used in the healing world, it is often in the green form, which links to the heart chakra. Some healers believe it can rekindle love, reduce bad tempers, and remove pollutants from the environment. Known as fairy treasure, green aventurine is thought to bring luck to those who place three stones in a dish in front of a gnome in the garden. Some healers believe green aventurine can help lower cholesterol.

Blue aventurine relates to the third eye chakra and is thought to help with self-discipline, including the desire to quit smoking, change unhealthful eating habits, and calm your spirits. Some healers suggest putting it under your pillow each night and being patient for the healing properties to take effect.

Red aventurine is related to the root, sacral, and third eye chakras and is said to foster focus and help you find your path.

AZURITE

— Chakras: Heart, Throat, Third Eye, Crown —

NAMING, LOCATION, AND HISTORY

This vibrant, deep blue-to-purple crystal has been found in Australia, Peru, Chile, Egypt, Russia, Morocco, and the U.S., especially in Arizona copper mines. Sphere-shaped azurite pieces have also been found in Utah.

Azurite is also sometimes called chessylite in reference to its discovery in the town of Chessy in eastern France.

Early civilizations ground azurite into a powder to use as a dye for clothing, and those who wore fabrics of that color were said to have high stature.

MYTHS AND LEGENDS

Ancient Chinese believed azurite would help them enter heaven, while Greeks and Romans thought the crystal would help them gain insight. The early Egyptians kept azurite from all but the high priests who knew how to tap into its powers. Native Americans considered azurite to be a sacred stone that would help them contact their spiritual guide.

USES IN HEALING

Today, crystal practitioners say azurite can be worn by students to help them learn, gain insight, and remember new facts. It's been said to relieve headaches, dizziness, and ringing in the ears.

Touching or rubbing azurite helps release its energy, according to healers. It's also touted to be particularly useful to those who want to deepen their meditation or calm their mind and instill compassion. Some say azurite works best with the third eye and crown chakras, while others consider it more useful with the throat and heart chakras.

Bright light, heat, and air can take away the brilliance of this crystal's color. To protect its sheen, keep it in a cool, dark place.

BLOODSTONE

Chakras: Root, Heart

NAMING, LOCATION, AND HISTORY

Bloodstone has been discovered in India, China, Brazil, Australia, and the U.S. Also called heliotrope, this crystal has recently been discovered on the northwest coast of Scotland.

MYTHS AND LEGENDS

Ancient Greeks called heliotrope the sun stone. They placed it in front of the setting sun and believed it cast a blood red color on the sun's reflection. They also placed it in cold water and then on the body to transfer the sun's power, which was said to prevent injury and heal wounds. Ancient peoples also ground it to a powder and used it to remove snake venom from humans.

Some believed if you gazed at bloodstone, any eye ailment would be cured.

Ancient peoples also used bloodstone to decorate cups, vases, and statues.

Legends say when Christ's blood fell onto the green earth, it turned to stone, creating bloodstone.

PHYSICAL PROPERTIES

Bloodstone's base color is green, but the most stunning part of the stone is the inclusion of red spots.

USES IN HEALING

Not surprisingly, crystal healers today say bloodstone can help those with blood disorders, such as anemia. Bloodstone has also been used to instill courage and increase physical stamina, and it's said to improve the immune system.

Some healers suggest wearing or carrying bloodstone to help you make decisions and ward off negative thoughts. Those with insomnia might try placing a bowl of water with bloodstone in it at their bedside in hopes of getting a good night's sleep.

CALCITE

Chakras: All

NAMING, LOCATION, AND HISTORY

The name "calcite" comes from the Latin word, "calc," which means "lime," and refers to the limestone found in shells of marine creatures and plants. Some of them, including trilobites, became extinct 250 million years ago.

Calcite can be found all over the world. The largest single crystal of calcite was discovered in Iceland and weighed more than 200 tons.

One of the most interesting forms of calcite is found in New Mexico. Massive calcite deposits on a streambed were discovered in 2001 within a portion of a huge cave complex. That section is known as the Snowy River Cave. Its name refers to the calcite formation, which extends for 11 miles. Snowy River is likely the largest continuous cave formation in the United States.

PHYSICAL PROPERTIES

Calcite comes in many colors including green, pink, orange, amber, and red. Found in more than 800 forms, calcite is a major component of marbles.

USES IN HEALING

Green and orange calcites are among those most often used by crystal practitioners. High-quality deposits of green calcite—from light to dark green—are found in Mexico. Calcite is said to work with the heart chakra to instill forgiveness and compassion. It's also been touted to combat infections caused by bacteria and help lessen the pain of arthritis. Some healers say placing green calcite on the heart chakra reduces fever, while placing it on a burn, will relieve the pain and reduce inflamed areas of the body. Some healers also claim it can help reduce the signs of aging.

Orange calcite has been linked to the root, sacral, and solar plexus chakras. Some practitioners call it the stone of the mind because of its ability to help students memorizing lessons. They also believe using orange calcite can reduce depression and shyness, improve the libido, and encourage good decision-making.

Found in massive formations in Mexico, orange calcite has been said to help with the digestive and hormonal systems.

Other calcites include an opaque pink form, which comes from Peru. Related to the heart chakra, pink calcite is touted to help overcome grief as well as heal after surgery and injury. The transparent pink form of calcite is also related to the heart chakra and is said to support heart health and reduce stomach problems related to stress.

An unusual amber-colored calcite is known as the stellar beam and relates to the solar plexus, third eye, and crown chakras. It's sometimes called the dog-tooth calcite because the crystals are shaped like a dog's teeth. Others say the crystals are shaped like rocket ships.

Stellar beam calcite is a variety of the stone that has been found in Tennessee, Missouri, and Montana. Crystal practitioners use stellar beam calcite for meditation as well as to remove negative thoughts. Stellar beam is more related to increasing one's spirituality, rather than improving physical conditions.

CAVANSITE

Chakras: Third Eye, Crown

NAMING, LOCATION, AND HISTORY

Cavansite was discovered in the 1960s in Oregon, but it can also be found in India. A deep blue specimen of the stone has been found there, and crystal collectors often seek after cavansite from Indian mines.

USES IN HEALING

Crystalogists say cavansite can help writers and teachers communicate effectively with their readers and students. It's considered a transitional stone, helping lead you into the next stage of life, using the lessons you've learned to make the change smoothly.

When combined with stilbite, cavansite can help calm and bring inner peace. The two stones together are useful as a sleep aid, according to crystal practitioners.

Crystal practitioners say cavansite can dispel feelings

of hopelessness and encourage joy and optimism. It's also been used as a dream guide, helping you remember your dreams and write about them upon awakening.

Those suffering with migraines and other illnesses caused by stress can use cavansite to relieve the symptoms, according to crystal healers.

PHYSICAL PROPERTIES

Cavansite can be opaque or translucent, and its color ranges from sky blue to blue-green. It is typically found among crystals of stilbite, which is white-to-pink. The white backdrop of stilbite with deep blue cavansite specimens atop can be a lovely collector's item. Crystal shop owners often sell cavansite in this form.

CELESTITE

Chakras: Throat, Third Eye, Crown

NAMING, LOCATION, AND HISTORY

Celestite, or celestine, is often found growing inside a geode, and the geodes containing celestite are highly prized. Some of the finest deposits have been found in the U.S. and Madagascar. Workers digging the Erie Canal in New York have discovered deposits of celestite, and large blue crystals have also been found in Michigan and California. Other sources include Mexico, Italy, and Poland.

The largest geode in the world, which was discovered in Ohio, contains crystals of celestite weighing up to 300 pounds, or more than 680,000 carats. Gustav Heineman founded a winery at this location in the late 1800s, and while workers were digging a well there, they discovered that the interior walls were covered

with celestite. The geode became known as the Crystal Cave. It's now open to tourists who can view the celestite crystals as well as taste and purchase wine.

One story suggests the discovery of the cave helped the winery owner stay solvent during the Prohibition. Over the years, some of the celestite was harvested and ground into a powder to create fireworks, but visitors can still see some impressive celestite formations at the cave.

PHYSICAL PROPERTIES

Celestite comes in white, gray, blue, green, yellow, orange, and brown, as well as a colorless

variety. Ohio celestite is a rare variety, which can be whitish to blue-gray.

USES IN HEALING

Wearing this stone is said to install calm and hopefulness, and it can help foster deep meditation. Celestites are also used to remove infections and help with digestive problems and weight loss.

CITRINE

Chakras: Root, Sacral, Solar Plexus

NAMING, LOCATION, AND HISTORY

Citrine's name comes from the French word meaning "lemon," or from a Greek word meaning "citrus."

The natural form is found in the U.S., Brazil, South Africa, England, Madagascar, Spain, Russia, France, and Scotland. It's been used in jewelry since ancient times. Brazil likely has the largest supply of this crystal in its natural form.

PHYSICAL PROPERTIES

The natural form of citrine, also called gold topaz, is pale yellow.

Deeper shades of citrine are actually amethysts heated to at least 1,000 degrees Fahrenheit, which is often what you'll find when searching to purchase this crystal. Heat-treated citrine has a reddish tint not seen on the natural form.

The natural and heat-treated citrines

have similar as well as different uses in crystal healing.

MYTHS AND LEGENDS

Chinese legend says citrine was the stone of success, and emperors used it in hopes of expanding their intelligence. Some students in China today use citrine to help them get good grades on exams, while professors carry it with them in class to help during lectures. The Chinese also include citrine in their healing practices to bring about abundance.

Along with some types of quartz, citrine was said to have magical powers as an aid to stave off evil thoughts and snakes.

During the 17th century, citrine was placed on men's kilts as well as on their swords and daggers.

During the 1800s, Queen Victoria used citrine to adorn the summer home she shared with Prince Albert. During the art deco period of the 1930s and 1940s, citrine became a popular crystal for use in jewelry as well as in furniture and home design.

USES IN HEALING

Citrine is called the light maker. It's been said to help improve your financial situation, as long as you work hard to obtain what you want. Keeping citrine in your purse, a wallet, or some other place where you keep money is said to increase your wealth. It's also said to absorb negative energy and then release it to the ground. As such, healers say citrine is one of few crystals that needs to be cleansed after use.

Natural and heated citrine are touted to improve your creativity, both at home and at work. It is also said to increase energy and instill a sense of security. Both forms are also said

well as add stamina to the body when exercising and performing physically demanding tasks. Healers also say it can help those who suffer from chronic fatigue syndrome.

Heated citrine is said to bring optimism into your life, help with weight loss, and help to maintain energy during exercise.

Crystal healers suggest wearing true citrine in a gold setting on a pendant with the pointed end facing downward toward the root chakra. Wearing citrine rings and bracelets keeps the crystal close to the solar plexus chakra.

to help heal afflictions of the heart, kidney, liver, and digestive tract.

Natural citrine is said to help find solutions to difficult situations, as

CLEAR QUARTZ

Chakras: All

NAMING, LOCATION, AND HISTORY

Quartz occurs worldwide and is sometimes referred to as rock crystal. When the word "quartz" is used, people often think of the clear quartz. You'll find other types of quartz with different properties in this book. See the entries for rose quartz, rutilated quartz, smoky quartz, and tourmalinated quartz—each with its own set of properties, colors, and relations to the chakras.

Other members of the quartz family in this book include amethyst and citrine.

Clear quartz can be found in the U.S., Brazil, Madagascar, and other countries.

MYTHS AND LEGENDS

Clear quartz has been named the perfect jewel by the Japanese, who thought it represented purity, patience, and perseverance. They believed the largest most brilliant clear quartz was a dragon's saliva, while the smaller ones came from the breath of a dragon.

Hindu mythology says the god Maya owned a crystal tank emblazoned with precious stones and pearls. People were lured in by transparency of the crystal, believing the quartz was clear water where they could bathe.

Ancient South American civilizations sculpted human skulls using rock crystal, and wealthy ancient Romans carried rock quartz in their hands to remain cool during heat spells.

Ancient priests used quartz to destroy magic and reverse spells.

Native Americans considered clear quartz a living being. Cherokee Indians used quartz to help them when hunting for food. They often hid their special rock crystals in caves or in their clothing.

Scotsmen and Irishmen once used clear quartz to rid their cattle of disease. Over many years, clear quartz has also been used to make crystal balls.

Many ancient cultures believed clear quartz could reduce pain, thirst, and fever, as well as cure colic. In addition, people of many different cultures over the years wore clear quartz crystals to obtain magical powers and induce rain.

PHYSICAL PROPERTIES

Clear quartz comes in a vast array of forms, including a long wand shape, a tabular shape, and one with points at either end.

USES IN HEALING

Considered a stone of life, clear quartz has been said to improve one's psychic abilities and increase certain emotions. Healers have said clear quartz can help hair and fingernails grow.

Crystal practitioners believe clear quartz can heal a wide variety of ailments including migraines, vertigo, fatigue, diarrhea, and digestive and skin disorders. It's also touted to help those trying to lose weight.

Clear quartz is also deemed to be

especially useful with the root chakra and when seeking to balance all the chakras. Clear quartz obtained from Madagascar has been said to be useful during meditation.

Placing a clear quartz crystal in a window is said to bring the sun's healing light and warmth indoors. It's also used with other crystals to improve their spiritual and physical healing powers.

Rock crystal is also known as the transformer crystal for its supposed ability to help improve health, relationships, and difficult sensations. In short, it is said to transform lives.

COVELLITE

Chakras: All

NAMING, LOCATION, AND HISTORY

Covellite was discovered in the 1800s at Mount Vesuvius by mineralogist Niccolo Covelli. It's often associated with copper mines, and it can be found in Italy, the United States, and Peru, among other countries.

PHYSICAL PROPERTIES

Covellite's color ranges from deep blue to deep purple or black.

MYTHS AND LEGENDS

Covellite has been said to facilitate the journey into one's Akashic records—a collection of thoughts, words, and experiences in past, present, and future lives.

USES IN HEALING

Practitioners suggest using covellite as a way to gain guidance regarding an issue in your life. One way to use covellite, according to crystal therapists, is to sit quietly in a dim room with the stone in your hands while thinking of the issue you want to solve. Looking at the stone's surface and seeing its sheen may bring images that help offer answers, according to one crystal therapist.

Covellite is touted as being helpful with women's reproductive issues. It's also considered a stone useful in transforming negative aspects of a relationship into positive ones.

Pairing covellite with turquoise is said to help make dreams come true by providing you with a positive attitude. Its high copper content is also said to help those with arthritis and rheumatism.

DANBURITE

Chakras: Heart, Crown

NAMING, LOCATION, AND HISTORY

Found in the U.S., Russia, Japan, Madagascar, Italy, and other countries, danburite was first discovered in Danbury, Connecticut. However, the largest quantities of the stone are found in Mexico.

PHYSICAL PROPERTIES

Danburite is pale yellow to colorless, and some crystalogists have claimed to see a Buddha design within the crystal.

USES IN HEALING

This is considered a crystal of laughter and joy, helping you change your mood even when surrounded by negative people. It's also said to help release anxiety and stress. Practitioners suggest holding danburite in both hands to experience calm or placing it beneath the pillow at night to promote deep and restful sleep.

Sitting barefoot outdoors and holding danburite to the sun has

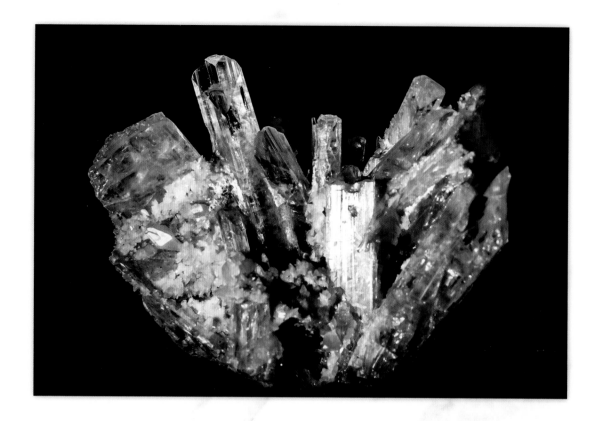

been said to imbue you with joy and serenity. Practitioners also say that carrying danburite in your pocket can help you remain calm during troubling situations.

A pink variety of danburite is considered especially helpful during meditation. A golden-yellow variety of danburite has been used to help remember dreams as well as to reduce the symptoms of allergies. Combining it with shungite is said to help intensify allergy-reduction.

Twin danburite crystals can help heal relationship issues, according to some crystal therapists.

Combining danburite with aquamarine, cavansite, and diopside is said to help those dealing with grief.

DIOPTASE

Chakra: Heart

NAMING, LOCATION, AND HISTORY

Formerly called copper emerald, dioptase has been found in Chile, Peru, the U.S., Russia, and several countries in Africa. It is usually found in or near deserts.

Its name comes from Greek words meaning "to see," which is likely in reference to its transparency. Plaster statues found in Jordan dating to 7200 BC are likely the oldest evidence of man's use of dioptase. The green around the eyes of three human figures called Micah, Heifa, and Noah were made of dioptase.

In the late 18th century, workers at a copper mine in Kazakhstan thought they had discovered huge emerald deposits within quartz. An analysis later showed the crystals were dioptase, just as beautiful as emerald, but not nearly as hard.

PHYSICAL PROPERTIES

Dioptase is transparent or translucent, and its color ranges from deep blue to emerald green.

USES IN HEALING

Some crystalogists say you can feel

dioptase vibrate when held to your heart chakra.

It is said to help resolve emotions stuck inside your inner child, like jealousy and anger. This helps you release the past and be free to live a life of abundance.

Healing properties cited for dioptase include relieving migraines, improving the immune system, and restoring peace. Using dioptase with stichtite and rhodonoite has been touted to reduce depression and fear.

Overall, working with this stone can bring peace and joy and relieve physical issues related to past trauma, according to crystal healers.

EMERALD

Chakra: Heart

NAMING, LOCATION, AND HISTORY

Emerald was first used by the Egyptians about 4,000 years ago. The most exquisite emeralds are found in Colombia, but they have also been discovered in Russia, Africa, and Brazil.

Emeralds are likely two billion years old. Humans began synthesizing emeralds in the 1960s. These can often be identified by their lack of imperfections.

Many emeralds of gemstone quality have been recorded throughout the ages. A stone purportedly weighing up to 840 pounds—though some say it's only 752—was discovered in Brazil in 2001, and is said to contain the largest single shard of emerald in the world. Conflicts over its rightful owner have been debated in court, and it's been stolen at least once. It's been valued between $75 million and $400 million. *National Geographic* even created a documentary about this crystal entitled *$400 Million Emerald Mystery.*

To further cloud matters, another huge emerald was discovered near the same location in 2017 and is said to weigh 794 pounds and be valued at more than $300 million. It took 10 people to extract the emerald from the mine where it was found. Both these emeralds are referred to as Bahia emeralds. Bahia is the state in Brazil where the mine is located.

Two much smaller emeralds, weighing about one pound each and found in North Carolina, have been displayed in the American Museum of Natural History. Others of similar sizes are at different museums, and the hope is that the two Bahia emeralds will eventually be housed in museums for the public to view.

MYTHS AND LEGENDS

Legend says the key to enlightenment is written on an ancient tablet made of emerald.

Ancient Greeks treated emeralds with cedar oil to enhance their beauty. Other ancient peoples

believed they could ward off evil spirits, and if someone put an emerald under his tongue, he would be compelled to tell the truth. Crystal practitioners today think combining emerald with aquamarine can encourage truth.

Incans in South America built small mines to harvest emeralds. When the Spanish took over, they built large mines and forced the Incans to dig them, according to some historical accounts.

PHYSICAL PROPERTIES

Emerald is the green form of the mineral beryl. Its color ranges from soft to deep green.

USES IN HEALING

Emeralds are said to instill unconditional love as well as compassion, understanding, and acceptance of others. Overall, emerald is considered a healer of emotional imbalances as well as of diseases of the heart.

Crystal healers suggest putting an emerald on the heart chakra to instill calm. Some recommend placing six emeralds on the various chakras for the best benefits.

Emerald has been called the stone of successful love when related to romantic matters. It is said to rekindle romances as well as help you discover a new love.

FLUORITE

Chakras: All

NAMING, LOCATION, AND HISTORY

Fluorite can be found in Germany, England, Spain, Canada, and the United States.

It arose 150 million years ago when hot water containing the chemical fluorine rose from the earth's interior to the surface. It crystallized in calcium rocks near fault limestone beds in the Midwestern United States. Native Americans used fluorite to fashion beads and other items.

A deep green fluorite found in New Hampshire is linked to the heart chakra. Yellow fluorite from Argentina is said to be in line with intellect, while blue fluorite is related to the throat chakra.

A blue fluorite called Blue John is only found near Derbyshire, England, and it

includes bands of colors including grey, yellow, blue, and purple. Those colors likely came from the mineral's contact with oil millions of years ago.

PHYSICAL PROPERTIES

Fluorite comes in blue, green, yellow, and other colors, but purple is the most common and recognizable. Several colors can exist in one crystal. When displayed under ultraviolet light, some fluorite crystals glow, giving off what's called fluorescence.

MYTHS AND LEGENDS

Folklore says rainbows and magic come from fluorite.

USES IN HEALING

Crystal healers link fluorite to a clear mind, and the crystal is said to help reduce conflict and aid in concentration. Called the brain stone, fluorite is said to improve memory and lessen vertigo.

Crystal therapists recommend using blue fluorite to help with eyes, nose, ear, and throat issues and using green fluorite for stomach disorders. Yellow fluorite is touted to lower cholesterol, while pink fluorite is said to relieve headaches.

GARNET

Chakras: All

NAMING, LOCATION, AND HISTORY

The name "garnet" comes from a Latin word meaning "pomegranate," referring to its color, which resembles the seeds of the fruit.

This long-lasting stone was discovered in a beaded necklace in a grave of man who died sometime around 3000 BC.

Travelers visit castles in the Czech Republic to view the interiors decorated with garnet and to purchase the stone. Garnets were marketed in Prague during the 17th century and the industry attracted many prospectors from Venice searching for the gem.

MYTHS AND LEGENDS

Ancient Asian peoples used garnets in their weaponry, thinking it would help them more easily wound their enemies, while others who wore it thought it would protect them from being wounded.

A single light found on Noah's Ark came from a garnet, according to the Talmud, a collection of Jewish writings.

Jewelers during the Middle Ages used garnet in rings, necklaces, and pins, and people wore these pieces to protect themselves against negative energy.

PHYSICAL PROPERTIES

A wide variety of colors besides red exist in garnet, and some are linked to different chakras.

USES IN HEALING

All garnets are linked to pleasure and prosperity and can lead to a sense of grounding, according to practitioners.

Different garnets and their chakra properties follow:

Spessartine garnet is yellow-orange and is named after the forested mountainous region in Germany where it was discovered. Spessartine garnet is linked to the root, sacral, and solar plexus chakras. It's said to unblock creativity as well as inspire confidence and optimism. It's also been used to help with weight loss.

Rhodolite garnet ranges from a rose to violet color and has been found in Brazil, Tanzania, Madagascar, Norway, the U.S., and other countries. Related to the root, heart, and crown chakras, rhodolite is considered a stone that can help lead you to your purpose. In addition, practitioners say it can treat heart and lung disorders and can help those recovering from feelings of shame and guilt.

Almandine garnet is linked to the root chakra and comes in hues of orange-red to purple-red. It's found in Brazil, India, Madagascar, and the U.S., with smaller amounts discovered in Austria and the Czech Republic. Sri Lankan almandine garnet is sometimes called Ceylon ruby.

Practitioners say it can help when recuperating from surgery as well as reduce anxiety, especially when it comes to finances.

Rainbow garnet, related to the heart chakra, comes from Africa and is found in red, green, and blue-green hues. It's considered a happy stone, one that can encourage a smile on your face. Practitioners say rainbow garnet makes an excellent host gift and can help keep a heart strong.

Black andradite garnet works with the root chakra. The deep purple-black crystal is found in Greenland and Mexico. It's said to help support the immune system and lead people on their true spiritual path.

Andradite also comes in yellow and green shades. Green andradite is said to reduce loneliness, and the rare yellow variety works with the solar plexus chakra.

HELIODOR

Chakras: Solar Plexus

NAMING, LOCATION, AND HISTORY

Heliodor's name comes from two Greek words meaning "sun" and "gift," and those who discovered it in Namibia in the early 1900s considered it a gift of the sun.

It's found today in Namibia, Brazil, Sri Lanka, Madagascar, France, and other countries.

PHYSICAL PROPERTIES

Heliodor is a variety of the mineral beryl, and its golden-yellow color resembles that of a ray of sunshine.

USES IN HEALING

The stone is said to bring out assertiveness, confidence, power, and strength, especially when it comes to fulfilling dreams. Crystalogists also have said heliodor is a good crystal for leaders to use because it helps them be firm but compassionate. Heliodor is said to help with intestinal distress and indigestion.

Wearing heliodor as a pendant is suggested to bring out its traits and to offer energy amid lethargy and fatigue. The stone is said to balance the solar plexus chakra, located above the navel and below the rib cage, helping boost the immune system and dispel fears related to having to please others.

Be careful when choosing heliodor—disreputable dealers irradiate aquamarine crystals to make them look like heliodor.

Some crystal practitioners suggest using heliodor to add warmth to the body during cold winters and to help those who suffer from seasonal affective disorder (SAD), which can cause depression.

IOLITE

Chakras: Solar Plexus, Third Eye

NAMING, LOCATION, AND HISTORY

Iolite can be found in Brazil, Madagascar, Myanmar, India, and Sri Lanka. It comes in various shades of violet-blue, so it has been nicknamed "water sapphire." The name "iolite" likely comes from a Greek word meaning "violet."

MYTHS AND LEGENDS

Legend says that from about AD 790 to 1066, Viking explorers used iolite as a way to view the sun, know their location, and determine their way back home. It's been called the Viking compass stone because of this.

PHYSICAL PROPERTIES

When viewed in different positions, iolite may change colors from blue to dark violet to light blue to even yellow or gray.

USES IN HEALING

Crystal practitioners suggest using iolite to help solve extreme emotional issues by fostering a clear mind, even when faced with major challenges. It's also said to help cure

sleeping disorders and maintain healthy eyes and a good memory. Practitioners say it can instill empathy and increase the ability to decide if someone needs help.

Iolite can also help in decision-making. Practitioners suggest placing an iolite stone on the brow and holding one in each hand in order to tackle difficult decisions.

A naturally occurring mix of iolite and sunstone is called iolite-sunstone. Its color is a deep rich purple to purple-black. Found in India, it's said to simulate the solar plexus and third eye chakras, just like iolite and sunstone do alone. Iolite-sunstone has been said to help with weight loss and combating stage fright.

JADE

Chakras: Root, Solar Plexus, Heart, Third Eye, Crown

NAMING, LOCATION, AND HISTORY

Jade comes in two forms—nephrite and jadeite. Nephrite is found in China, New Zealand, and Russia. Dark green jadeite comes mainly from North America, while light green jadeite has been discovered in Russia, China, and Guatemala.

The name "jade" has a peculiar origin. It comes from a Spanish phrase for "a stone used to treat pain in the side," because Spanish explorers encountered people who used jade at their sides to relieve pain. Other accounts translate the phrase to mean "kidney stones," as explorers believed it could be used to relieve kidney ailments, as well.

Since ancient times, the Chinese have revered jade, using it to make weapons, tools, and carvings. They considered it to be a source of luck, a pure spirit, and a clear mind.

Today, they continue to carve nephrite and jadeite into animals and symbols. These include butterflies, which symbolize a long life, and a

flat, circular object with a hole in the center, signifying heaven.

Chinese jade ornaments are not all green; a white and brown jade ornament portraying flowers and grape leaves from the Jin dynasty, which thrived in the 12th and 13th centuries, can be found at the Shanghai Museum. A 12th-century cup with dragon handles is made of light green jade and housed in another museum in China.

PHYSICAL PROPERTIES

Jade can be found in a host of colors, including green—the color with which it's most identified—as well as white, purple, black, red, lavender, and light blue. Each color has its own relation to a different chakra. Nephrite is typically white or various shades of green, while jadeite can be blue, red, dark green, lavender, or white.

USES IN HEALING

Overall, jade is said to balance the heart chakra, offering relaxation to the user as well as the ability to improve relationships. It's also linked to longevity.

Jade in shades of green is said to bring joy back to your life.

Green nephrite jade has been touted as a way to strengthen the heart and the nervous system, while green jadeite has been used to speed healing after surgery.

Black jade is considered a protective stone that can deter those who want to harm you either physically or mentally. Healers also say it can protect the immune system—especially when traveling—and prevent viruses and bacterial infections. It's linked to the root chakra, and it is said to dispel negative emotions.

Blue jade, related to the third eye and crown chakras, is considered the calming stone. Blue jade, a variety of jadeite, is said to offer a soothing spirit, helping you remain calm during difficult situations. It's also touted to help with arthritis, asthma, and bronchial viruses.

Purple jade works with the third eye and crown chakras, according to crystal practitioners, and helps restore joy amid negative feelings. It can also help you discern the truth. It's been said to calm the nervous system as well as relieve hives and rashes related to stress.

Red jade focuses on the root and solar plexus chakras, and is said to help those who need to assert themselves and those trying to overcome addiction. Some have used red jade to increase their financial prosperity. It can also be used to help people realize that they have everything within themselves that they need to be happy and content.

KUNZITE

Chakras: Heart

NAMING, LOCATION, AND HISTORY

Kunzite was first catalogued in 1902 by jeweler George Frederick Kunz, the stone's namesake. It can be found in the U.S., Brazil, Australia, Afghanistan, Myanmar, and Madagascar.

PHYSICAL PROPERTIES

Known as a gentle stone, kunzite is a transparent pink to light purple form of spodumene. Some gem dealers enhance this color with irradiation and heat. Its color can fade when exposed to heat and light, so keep it in a closed container when not using.

USES IN HEALING

Related to the heart chakra, kunzite is said to engender love of the self, humanity, nature, and the divine spirit. Practitioners suggest offering kunzite as a gift to others to encourage their physical and emotional health.

Some crystalogists suggest wearing or carrying kunzite when going through major changes, such as a new job, a relationship breakup, or retirement. The crystal is said to ease the transition and keep you positive.

Kunzite is said to be soothing and calming, making it useful when taking a test or going on an interview for a job, and helping to curb anxiety. It's touted as a protector against road rage and the stress of driving.

Kunzite has also been linked to the reduction of rashes due to allergies, the strengthening of the heart and circulatory system, and the easing of depression and other mental disorders.

LABRADORITE

Chakras: All

NAMING, LOCATION, AND HISTORY

In 1770, in eastern Canada's Labrador Peninsula, missionaries found a blue-gray crystal showing iridescent green and blue hues. They named it labradorite, after its discovery place. This vividly-hued crystal is also found in Madagascar, Finland, and other parts of the world. The term *labradorescence* was coined to describe the optical effect of the iridescence of the stone, and has been explained by a particular type of structure. Depending on how you turn the stone, different flashes of colors will appear.

A rare variety of labradorite called spectrolite was found in the 1940s in

Finland. It displays a wide range of colors including blue, green, yellow, orange, and red. Finns began mining spectrolite to create jewelry after World War II.

MYTHS AND LEGENDS

Native Americans believed that the crystal reflected the Aurora Borealis, a natural light show of greens, blues, and other hues seen mostly in the far north. Ancient Inuit legend says labradorite fell from the Aurora Borealis to the earth.

Another Inuit legend says the Aurora Borealis was trapped in rocks along Labrador's shoreline and that a warrior speared the ground trying to free it. The warrior was unable to

get all of the crystals and left behind those that glimmered in the light.

Other myths purport that alien beings from the sky are trapped within the stones. People today say this stone is a reminder to go outside and gaze at the sky.

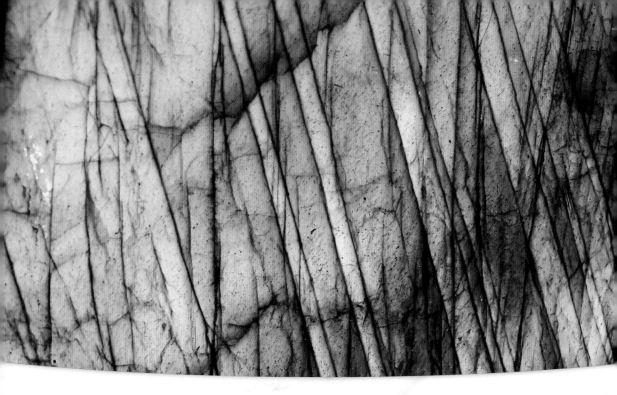

USES IN HEALING

Wearing labradorite is said to help you maintain your energy when dealing with relationship problems, eventually leading to a better balance between you and your loved ones.

Practitioners suggest releasing negative energy from the body by placing labradorite stones in key places: one at the head, one at each foot, and one at each side of the body. Placing a fluorite crystal at the throat is said to enhance this effect.

Some other practitioners claim labradorite can heal disorders of the brain and eyes, lower blood pressure, and relieve pain.

Others say labradorite particularly relates to the throat chakra which, if blocked, takes the rest of the chakras out of balance. Labradorite is said to open the throat chakra to enable you to understand and reveal your truth, communicating more effectively to others. Labradorite with lighter shades of blue is said to instill relaxation.

Holding labradorite is said to bring you closer to the mystical powers of the northern lights and to bring about positive changes in your life.

LAPIS LAZULI

Chakras: Throat, Third Eye

NAMING, LOCATION, AND HISTORY

Lapis lazuli's name comes from the Persian word for "blue." Deep blue varieties of lapis are found in Afghanistan. The crystal has also been discovered in Russia and China, and smaller amounts have been found in the U.S., Italy, and India.

The brilliant blue color of lapis lazuli has drawn ancient cultures to harvest it and fashion it into jewelry for centuries. One set of lapis lazuli beads dates to 700 BC.

During the Renaissance and baroque eras, artists crushed lapis lazuli to create a pigment called ultramarine for their oil paintings. They used the pigment for painting the clothing of the Virgin Mary and Jesus Christ.

This pigment was extremely expensive. One story says Michelangelo left part of one of his paintings unfinished because he could not afford the ultramarine pigment he wanted to use to complete it. Another story says Dutch painter Johannes Vermeer, who lived during the 1600s, went into debt from his purchase of the prized pigment.

In the early 19th century, a less expensive synthetic form of the pigment was created, and artists began using that instead.

Some painters, including Wassily Kandinsky, continued to use ultramarine throughout the 20th century because they believed it had spiritual powers.

MYTHS AND LEGENDS

Ancient Egyptian kings and queens were buried with lapis lazuli, and King Tut's stone coffin contains inlays of the stone. It was thought to help with the journey into the afterlife. Ancient Egyptians also believed lapis lazuli purified the soul and healed mental illness, and that by grinding it, mixing it with gold, and placing it on the head, it could remove one's demons. Cleopatra was said to use powder made from the crystal as eye shadow.

During the Middle Ages, lapis lazuli was considered to be the blue of heaven, and it was believed that the crystal could banish dark spirits and bring in wisdom and light. Buddhists considered lapis a stone of peace. Catherine the Great, who lived during the Renaissance period, decorated the doors, walls, and fireplaces in one room in her palace with lapis lazuli.

PHYSICAL PROPERTIES

Lapis lazuli consists of a rock of blue lazurite containing calcite, pyrite, and other minerals.

USES IN HEALING

Lapis lazuli is known as a stone of truth, harmony, and friendship.

Crystal practitioners believe lapis lazuli connects to the third eye, helping people seeking spiritual guidance. It's also considered a stone to help increase intelligence and knowledge as well as improve memory. Some have said journalists, lawyers, and psychologists should use lapis lazuli to improve their

senses of wisdom and judgment.

Lapis lazuli, along with turquoise and other stones, has also been touted to relieve migraines when placed on the forehead. Some crystal practitioners have said lapis lazuli can boost the immune system, lower blood pressure, cure insomnia, and relieve depression.

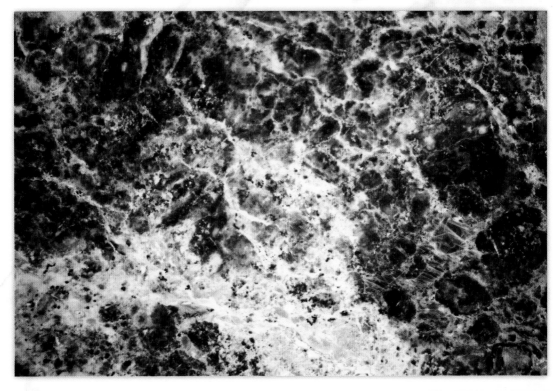

LARIMAR

Chakras: Throat

NAMING, LOCATION, AND HISTORY

According to records, larimar was likely first discovered in the early 1900s, but the explorer who found it wasn't allowed to retrieve the stone. In the 1970s, larimar was rediscovered, and it was learned that native people in the Dominican Republic called it a stone from the sea. The stone was named after the daughter of Miguel Mendez, Larissa, and the Spanish word for "sea," which is "mar." The name refers to the hues of the Caribbean Sea. Larimar is sold on the island in jewelry.

It is only found in the Dominican Republic on the Caribbean island of Hispaniola.

PHYSICAL PROPERTIES

Larimar forms within lava and often contains reddish-brown inclusions within its main colors of blue and

white. It's considered a blue variety of pectolite. Too much light and heat can cause the blue color of larimar to fade.

USES IN HEALING

Those who wear larimar are said to be surrounded with a sense of tranquility and are able to communicate with strength and an open heart. The crystal is also said to help protect a singer's voice.

Some crystal practitioners believe larimar can help those who suffer from panic attacks and fear. They also say that it can help lower blood pressure and reduce infections and fever. It's been called a worry stone, and those who rub the polished crystal may be able to reduce their anxiety.

Larimar is also touted as a useful cold and flu remedy.

LEPIDOLITE

Chakras: Heart, Third Eye

NAMING, LOCATION, AND HISTORY

Lepidolite has been discovered in the United States, Russia, Brazil, Madagascar, and Canada. A lilac lepidolite is found in Zimbabwe. It's been mined extensively in the New England states.

Its name comes from a Greek word meaning "scale," referring to the scaly flakes of lithium within the stone. Lepidolite is a major source of lithium, one of the world's lightest metals, which is used in rechargeable batteries, mobile phones, and mood-stabilizing pharmaceuticals. Lepidolite's touted ability to relieve stress is linked to its lithium content.

PHYSICAL PROPERTIES

Typically pink to lavender, lepidolite can also be gray, white, or colorless.

USES IN HEALING

Lepidolite has recently been discovered by crystal healers, and it is known as the new age stone.

It's also known as a soothing stone, and has been touted to help those with insomnia and nightmares due to stressful situations and relationships. Carrying it with you is said to help you take a breath and get centered before reacting to difficulties. Practitioners have also recommended those with arthritis and similar issues to use lepidolite.

The variety lilac lepidolite is indeed lilac, but can also be found in pink, purple, and lavender. Practitioners say lilac lepidolite increases one's ability to think analytically as well as to feel a sense of peace about life. Some also say it can help those who have had an emotional breakdown recover as well as aid those dealing with difficult changes such as a relationship breakup or job loss.

MALACHITE

Chakras: Solar Plexus, Heart

NAMING, LOCATION, AND HISTORY

Malachite's name may have come from a Greek word for a green herb whose color resembled that of the stone. The stone can be found in The Democratic Republic of the Congo, Russia, and the United States, among other countries.

MYTHS AND LEGENDS

Malachite was known as a protective stone to Ancient Egyptians, who created carvings of this brilliant green stone often set with lighter green bands. They placed malachite carvings near children to protect them, and hieroglyphics from the time period revealed malachite as a

testament to the presence of gods. Pharaohs also placed malachite inside their headdresses, believing that the stone would make them sage rulers.

In Italy, people wore malachite amulets called peacock stones to protect them from evil. During the Middle Ages, people thought malachite would stop fainting and prevent hernia. When ground into a powder, it was used to treat muscle aches, and when mixed with honey, it was said to stop wounds from bleeding.

USES IN HEALING

Stone therapists today think malachite will protect people from radiation caused by computers and microwaves and that it can treat those with asthma, arthritis, and digestive problems. Holding malachite over areas of the body experiencing pain after long exposure to computers is said to absorb the radiation. In addition, it's touted as a stone that can help reduce confusion and turn creative thoughts into actions.

MOLDAVITE

Chakras: Heart, Third Eye

NAMING, LOCATION, AND HISTORY

Some 15 million years ago, a meteorite hit the earth somewhere in what is now the Czech Republic, resulting in the formation of moldavite. Scientists speculate whether this stone was part of the meteor itself, or if it formed when it reached the earth.

Moldavite is thought to be a natural glass rather than a crystal. Over the years, farmers in the region where the meteorite fell found pieces of moldavite in their fields.

PHYSICAL PROPERTIES

Moldavite can be found in several shades of green or blue-green. It can be transparent or translucent and have a rough surface that makes it look like moss.

MYTHS AND LEGENDS

Since ancient times, moldavite has been the stuff of legend. Neolithic people from 25,000 BC may have made jewelry from moldavite, which they wore to bring good fortune.

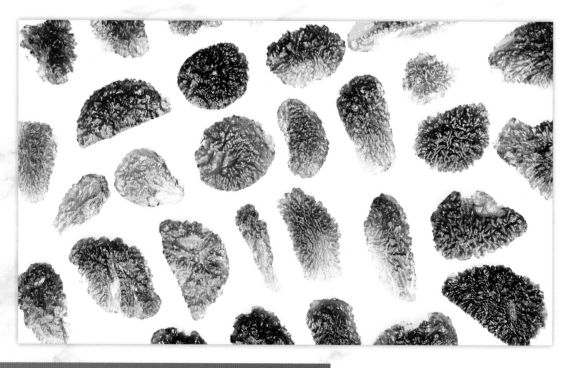

Legends also say when the archangel Michael sliced his sword at the crown of the devil Lucifer, a green stone from the crown fell to earth. The stone was later said to be made into a cup, called the Holy Grail, used at the Last Supper of Jesus Christ.

During the legendary King Arthur times, men supposedly drank from a Holy Grail made of moldavite to heal and to remain true to their loves.

USES IN HEALING

Crystal practitioners say moldavite can be used during meditation to heal and enhance relationships. They say it can be highly transforming and those who use it should be prepared for strong experiences. Moldavite is considered a spiritual stone above all, rather than one that heals specific physical ailments.

MOONSTONE

Chakras: Sacral, Third Eye, Crown

NAMING, LOCATION, AND HISTORY

Moonstone is aptly named for its pearly, light blue luster as well as ancient people's beliefs that placing the stone on their tongue under a full moon would make them clairvoyant.

It is found in Sri Lanka, India, Madagascar, and other countries, including the United States.

Ancient peoples from India would give moonstone as a wedding gift to forecast a harmonious marriage. They also believed that it was a stone of fertility and passion.

Today, people of India consider moonstone to be the best stone for women to wear. It's additionally used with Ayurvedic medicine in India, which also involves herbs, diet, and massage.

Moonstone has been used in Roman jewelry for at least 2,000 years, and ancient Romans believed that drops of moonlight formed the crystal.

Florida named moonstone its state gem in 1970 to honor the U.S. Kennedy Space Center in Brevard

County and man's landing on the moon in 1969; however, the crystal is not found in Florida or on the moon.

MYTHS AND LEGENDS

Asian legend says the tides bring blue moonstones to shore every 21 years and that hanging it on fruit trees will guarantee excellent crops.

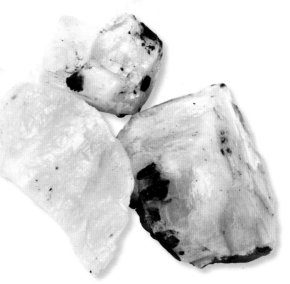

Another legend says a fight between two gods resulted into one of them falling to the earth, and that his sparkling eyes became moonstone.

A 16th-century French astronomer supposedly experimented with moonstone and claimed that its appearance changed with the lunar cycle.

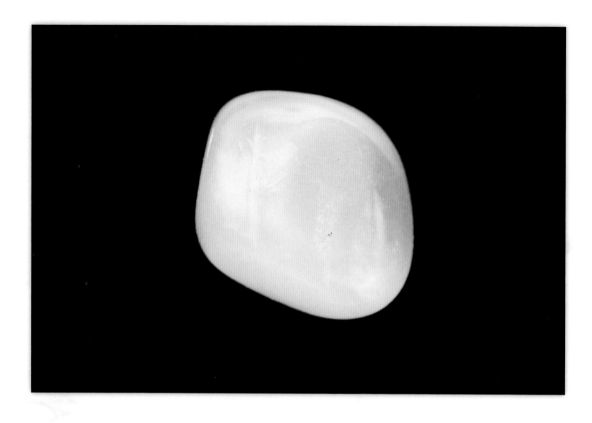

USES IN HEALING

Crystal practitioners say moonstone can balance a woman's menstrual cycle and help people be patient in seeing another person's point of view. One healer suggests placing one moonstone on the head, one on each shoulder, and one on each hip bone to relieve the mind and the body of stress. As a result, this relieves physical ailments associated with anxiety. It's also been said to help those who are grieving.

Moonstone is found in several different shades as well as in what's called a cat's eye. Each of these has different

attributes, according to crystal practitioners.

Cat's eye moonstone is a rare crystal that looks like a slit in a cat's eye when viewed a certain way. It's mostly found in Sri Lanka and India and is said to balance energies and help understand life lessons.

Gray moonstone is said to remind the user that all things are possible. White moonstone is touted to help men achieve emotional balance and to rid children of nightmares and insomnia. Peach moonstone—sometimes called yellow moonstone—reduces anxiety and encourages love, according to healers.

Rainbow moonstone is likely a type of white labradorite rather than a true moonstone, although not all crystal practitioners agree. It's touted to act as a prism, cleaning the mind and helping you sleep.

MUSCOVITE

Chakras: Third Eye, Crown

NAMING, LOCATION, AND HISTORY

Muscovite has been found in Russia, Austria, Switzerland, Brazil, and the U.S., among other countries. The name comes from the Muscovy region of Russia, where the stone was used to make glass for windows in ancient times.

Muscovite is the most common mineral in the mica family, and it is ground to use in the manufacturing of various products, including electronic devices and insulation.

PHYSICAL PROPERTIES

Muscovite is typically colorless, white, or gray, with tinges of other colors such as yellow, brown, blue, and pink.

MYTHS AND LEGENDS

Celtic civilizations believed yellow muscovite, known as star mica, was stardust and came from the twinkle in stars.

USES IN HEALING

Crystal practitioners say muscovite can be used as a tool for making important decisions or for helping you learn a foreign language and succeed at exams. It's also said to help regulate blood sugar levels, cure insomnia, and help those who are trying

to lose weight or are suffering with allergies.

Some practitioners believe using muscovite can help people overcome the end of a long relationship and help sensitive people not to overreact to their emotions. Those who are in tune to psychic awakening are encouraged by healers to use muscovite to prevent an overload of experiences that could lead to dizziness and headaches.

A dark pink muscovite known as red muscovite is said to bolster self-confidence as well as offer the healing properties muscovite of different colors provides.

OBSIDIAN

Chakra: Root

NAMING, LOCATION, AND HISTORY

Obsidian's name comes from a Roman explorer called Obsidius, who was said to have discovered obsidian in Ethiopia. Obsidian is actually a volcanic rock, and it was once used to make arrowheads by peoples living in the Stone Age at least one million years ago. It was once thought that the debris from meteorites fallen to the earth were obsidian. Today most scientists discredit that theory.

Obsidian has been found in places where certain types of volcanic eruptions have occurred. These include Argentina, Canada, Australia, Greece, Iceland, Italy, Kenya, Mexico, and the United States. Mountainside deposits of obsidian are found at Yellowstone National Park. You can also hike on obsidian flows at the Big Obsidian Flow

Trailhead and Interpretive Center in the Deschutes National Forest in Bend, Oregon.

Because of its ability to be fractured into sharp objects, obsidian has been used since ancient times to make blades and other tools. Archeologists discovered obsidian artifacts dating to the copper age of the Mediterranean region in about 3000 BC. The discovery showed trade routes of the ancient peoples. Tablets discovered on Easter Island dating back hundreds of years were carved with obsidian in what archaeologists believe were ancient writings.

Native Americans bartered with obsidian, and stories say they chipped flakes of the stone from mountains and then created crude blades, which they transported to different tribes, who then made them more stylized and useful.

Obsidian remains an important material today for making knives and scalpels for research animals.

PHYSICAL PROPERTIES

Obsidian is a dark, glossy black stone that has also been found with reddish brown, rainbow, or snowflake patterns.

USES IN HEALING

Crystal healers use several different types of obsidian, including black, which is said to help you know and accept your dark side and to help with eating disorders and addictions. It's also touted to balance the digestive system.

Snowflake obsidian is black with white patches made of mica and other minerals. Its healing properties are similar to those of black obsidian, according to practitioners. This variety, related to the root and third eye chakras, is also said to be calming and soothing.

Mahogany obsidian is a combination of colors, including black and reddish brown, reminiscent of mahogany wood. It is said to improve the function of the liver and kidney,

relieve pain, and instill a sense of worthiness.

Rainbow obsidian is polished black obsidian that reflects red, blue, gold, and green shades. Carrying rainbow obsidian is said to help remove negative thoughts, maintain a healthy heart during stressful times, and bring joy into your life.

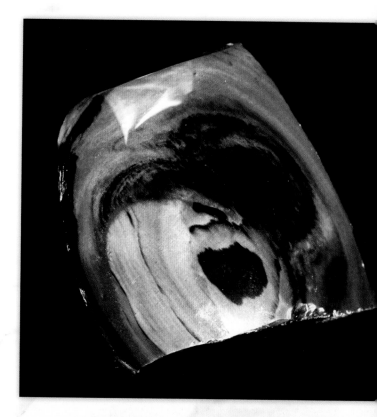

Peacock obsidian, which works with the root and third eye chakras, is fairly rare. When held in strong lighting, it shows red, gold, green, and blue patterns. Its properties are similar to those of rainbow obsidian.

Gold sheen obsidian, which relates to the root and solar plexus chakras, shows a golden sparkle when viewed at certain angles. It's said to help with digestive disorders like acid reflux disease, and it can help you find clarity in your life's purpose by revealing hidden talents.

OCEAN JASPER

Chakras: Solar Plexus, Heart, Throat

NAMING, LOCATION, AND HISTORY

This unusual variety of jasper is found only in the northwest coast of Madagascar. Deposits of ocean jasper are mined at low tide and are transported by boat.

MYTHS AND LEGENDS

Ocean jasper has been called the Atlantis stone because of its link to the fabled mythical island where it was said to have mystical powers.

PHYSICAL PROPERTIES

Ocean jasper varies in color from white to green to red to black, and it is often found with many patterns. Specimens of ocean jasper appear like white stones with different sized circles of green and orange embedded within.

USES IN HEALING

Ocean jasper is considered a stone of joy, able to help you express love and optimism. It is also useful for

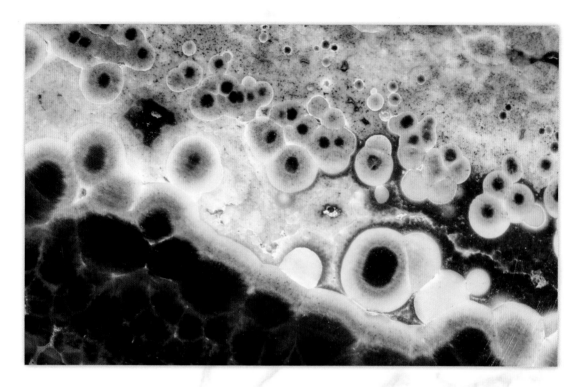

those who suffer from depression. Placing ocean jasper in work and living spaces is said to create a sense of relaxation. Healers say it can balance the body's chemistry and help with the absorption of vitamins and minerals.

Other healing uses include improving the digestive system and reducing skin disorders. Some have said ocean jasper can also help with seasickness.

Ocean jasper that includes a green color is touted to work with the heart chakra to help deal with the emotional ups-and-downs of relationships.

Those who believe in past lives use ocean jasper to journey back into earlier times and to reconnect with the earth.

Other jasper varieties used for crystal healing are listed in this book and are found in the U.S. and other countries. See picture jasper and unakite jasper.

ONYX

══ Chakras: Root, Solar Plexus, Third Eye ══

NAMING, LOCATION, AND HISTORY

The word "onyx" comes from a Greek word meaning "fingernail."

Onyx has long been used for carvings and jewelry. Ancient Greeks and Romans carved cameos from onyx. Art objects containing sardonyx—a naturally occurring orange variety—have been recovered from archaeological digs in ancient Greece, dating to circa 1400 BC. Ancient Egyptians used it to make bowls and other pottery. A green onyx from Brazil was used for sculptures and dishes. German artist Ferdinand Preiss used green onyx for many of his sculptures; and in Austria, green onyx has been used in dishes inlaid with bronze figures. Ancient Romans seeking courage carried sardonyx when they went to battle.

In Russia, the Mariinksy Theatre, which opened in 1860, features yellow onyx in the lobby. The Hotel de la Paiva, a huge mansion built in the 1850s in Paris, also uses yellow onyx in its décor, which includes a long winding staircase.

Famous architect Mies van der Rohe used slabs of onyx from Africa on a wall in the Barcelona Pavilion, which he was asked to design for an international exhibit in Spain in the early 1900s.

MYTHS AND LEGENDS

A Greek myth suggests onyx was created when the goddess Venus was sleeping on the banks of a river. Cupid used the point of one of his magical arrows to manicure her fingernails, which then fell into the river.

During Renaissance times, Europeans believed sardonyx improved one's speaking abilities.

Ancient Persians believed the stone could prevent evil and render the wearer invisible. Over the years, black onyx came to be equated with sadness and bad luck as well as a way to cool a lover's passion. In 1560, an astrologer wrote that onyx was used in India to stifle the libido. Chinese writings from the late 1800s mentioned that no one wanted to touch the stone, fearing they'd meet misfortune.

PHYSICAL PROPERTIES

Onyx, a variety of the mineral chalcedony, is dark with parallel bands of different colors, usually black and white. Some stones, such as agate, are dyed black and called onyx.

USES IN HEALING

Today, onyx is said to tackle your deepest fears. It's also said to help those with weak legs and is considered a stone of strength for those recovering from illness, embarking on difficult projects, or

those exhausted from physical exertion. Sardonyx, especially, is touted to keep a marriage happy and stable.

Healers suggest viewing onyx to quiet the emotions and bring about introspection, as well as provide guidance in understanding difficult situations. Emotionally, it's been said to instill self-control and reduce temper. Practitioners also say onyx can be used to focus when studying, learning, and working as an accountant or lawyer.

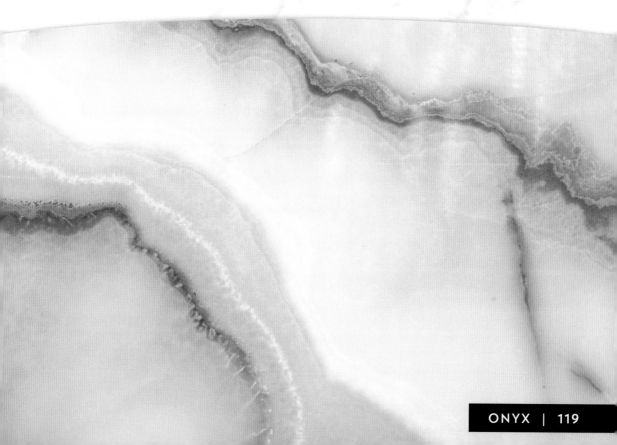

OPAL

Chakras: All

NAMING, LOCATION, AND HISTORY

Many types of opals exist, each with its own healing powers, according to crystal practitioners. Different types also come from different countries. For example, pineapple opal, which resembles its namesake, is found in Australia, while fire opal comes from Mexico.

Precious opal refers to those stones that show different shining colors when viewed through light.

Common opal is opaque and can be brown, black, pink, or blue. It can be found in Oregon and Peru, as well as other countries. Sometimes common opal is called "potch." The play of light does not affect common opal.

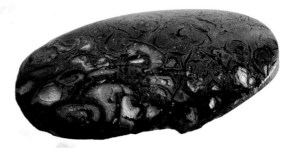

MYTHS AND LEGENDS

Over the centuries, opal has been linked to both good and bad luck and used by ancient civilizations in different ways. For example, the French once thought wearing opal could turn someone invisible, which would allow that person to steal.

Ancient Romans thought opal symbolized power. Venetians thought that opals could determine if someone had caught the plague. The opals would glow brighter if they had the plague, and then return to their original luster when the person died.

King Louis XIV of France was said to have named his coaches after crystals. One driver named Opal was usually drunk, linking the stone to bad luck.

USES IN HEALING

Pink common opal is said to instill a sense of peace and calm as well as help those with irregular heartbeats. Blue common opal is said to help you look upon the future with happiness, and it is also touted to help those with asthma, allergies, and psoriasis.

Brown or black common opal is used by practitioners to strengthen bones and help those with poor digestion.

One of the most commonly-used opals in jewelry is white precious opal, and it is said to relate to all chakras. Most white precious opals come from Australia, but others have been found in Guatemala, Japan, and the United States. This opal is said to strengthen the hair and fingernails as well as help those with eczema and rosacea.

Black precious opal is iridescent like white precious opal, but it has a dark

background. It's found in Australia and is said to work with the root and crown chakras. It's said to help overcome deep fears and anxiety.

Fire opal is mined in Mexico, the main source of this bright orange stone. Other places it has been found include Honduras, the U.S., and Australia. It's been used to help people with chronic fatigue syndrome and has also been said to awaken the libido and curiosity.

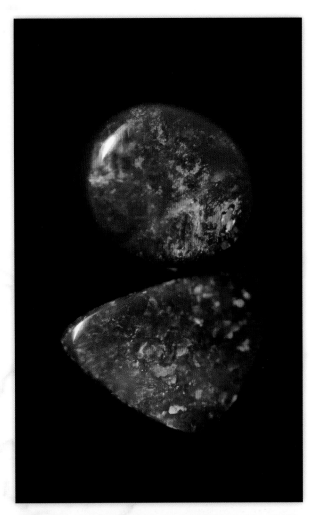

Another opal found in Mexico is violet flame opal. This stone, a combination of light purple, deep purple, and white, was discovered in 2011. It's linked to the heart, third eye, and crown chakras. Violet flame opal is thought to awaken spirituality and inner wisdom as well as bring harmony to the entire body.

Another new variety of opal was discovered in Oregon in 2003 and is called owyhee blue opal. It's said to reduce worry and calm throat and sinus infections and helpful in reducing inflammation.

PERIDOT

Chakras: Solar Plexus, Heart

NAMING, LOCATION, AND HISTORY

Peridot likely formed deeper inside the earth than other stones, and some meteorites that fell to the earth are even said to contain peridot. Olivine, the base form of peridot, was discovered on the moon between 2007 and 2009 by Japanese astronomers. Craters more than 50 miles in diameter contain large amounts of olivine, according to the astronomers.

The origin of the stone's name is uncertain, but an island where it was mined more than 3,000 years ago was named Zagbargad, which is Arabic for "peridot" or "olivine."

Part of the island is surrounded by the Red Sea, and Egyptians were said to have kept its location a secret. Peridot is known as the national gem of Egypt. It's also found in Pakistan, Brazil, South Africa, the U.S., and other countries.

MYTHS AND LEGENDS

Ancient Greeks considered peridot a symbol of the sun, which could bring energy to rulers who wore the crystal. Ancient Egyptians fashioned beads out of peridot. Legend says that the emeralds Cleopatra

wore were actually peridot, a stone she admired. One of the shrines within Germany's famous Cologne Cathedral was once thought to be decorated with emeralds. However, the stones are actually peridots.

Ancient peoples believed if they

pierced peridot, attached it to donkey hair, and wore it on their left arm, they would be protected from evil spirits. Placing peridot under the tongue was said to help those suffering with fever, and it was also used in powdered form to relieve asthma symptoms.

Ancient cultures also set peridot in gold and then wore it as a necklace or bracelet to protect them against evil spirits, nightmares, and mental illness, as well as to improve their intellect. People living in 13th-century England believed wearing peridot engraved with a torchbearer symbol would make them rich.

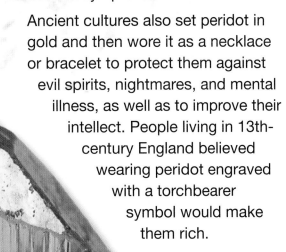

PHYSICAL PROPERTIES

Peridot comes in olive green, and the tint depends on how much iron is in the crystal. Some peridots can appear yellow to brownish-green.

USES IN HEALING

Crystal practitioners today believe peridot can generate spiritual and financial wealth as well as help people who believe in previous lives understand and make amends for past transgressions. It's also considered to be a stone to help those recovering from addictions as well as an aid to heal the heart. It's additionally known as an elixir for depression and helpful to those who are studying for exams.

One practitioner suggests using peridot to remove toxins from the body by placing stones on the throat and heart chakras and one near the kidney.

PICTURE JASPER

Chakras: Root, Third Eye

NAMING, LOCATION, AND HISTORY

Jasper's name comes from a Greek word meaning "spotted stone," and different varieties—including picture jasper—contain different colors and patterns.

Various types of jaspers can be found in the United States in California, Utah, Idaho, and Oregon, as well as other countries such as Australia, Brazil, Canada, Russia, Egypt, and Uruguay.

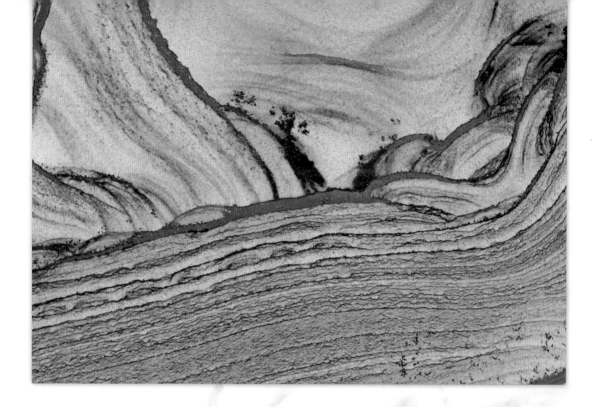

MYTHS AND LEGENGS

Ancient people and Native Americans used jasper to encourage rain, while others thought it could drive away evil spirits and keep spiders from biting.

PHYSICAL PROPERTIES

Picture jasper contains banded patterns of brown, gold, and gray, created when mud dripped into molten lava, and was then naturally heated and cooled. Picture jasper gets its name from its patterns that look like landscapes with mountains and valleys.

USES IN HEALING

Picture jasper is often used in meditation and has been said to encourage healing and growth in bones. Other practitioners use it to help those with digestive disorders, skin disorders, and allergies. Pairing it with lodestone is said to help those starting on new business ventures.

Practitioners also advise using picture jasper when traveling, quitting smoking, seeking inspiration when writing or creating art, and balancing the chakras. In addition, it's known as a comforting stone and one that can bring out ecological awareness.

PYROMORPHITE

Chakras: Solar Plexus, Heart

NAMING, LOCATION, AND HISTORY

First discovered in the mid-1700s, pyromorphite gets its name from the Greek words for "fire" and "change," referring to how the stone crystalizes when being cooled after melting.

Various shades of the crystal are found at the Bunker Hill mine in northern Idaho, where silver, zinc, and lead also exist. Collectors purchase expensive specimens of pyromorphite discovered in this mine.

Other places where pyromorphite has been found include Germany, France, Great Britain, and Australia.

PHYSICAL PROPERTIES

Pyromorphite is mostly green, but it can also be found in yellow, orange, brown, and gray. It has been described as looking like a branching cactus.

USES IN HEALING

Crystal healers link pyromorphite to the digestive system and say it can help maintain a healthy liver, pancreas, and gallbladder. It's been said to work well with other crystals to enhance their particular healing properties. It's also said to help alleviate symptoms when subject to toxic substances.

One healer suggest nature lovers can benefit from pyromorphite by bringing it with them on their outdoor adventures, including gardening and restoration work. Combining it with anthophyllite is said to connect humans to the spirits of nature.

Emotionally, pyromorphite is believed to clear negative energy and pave the way for a clearer future.

As it is related to the heart and solar plexus chakras, pyromorphite is also touted to help balance the mind and the heart.

PYRITE

Chakras: Solar Plexus

NAMING, LOCATION, AND HISTORY

Often called "fool's gold" because of its metallic luster, "pyrite" comes from a Greek word meaning "fire." That's likely because when striking it against stone or metal, typically iron, it produces sparks. Pyrite has been used to make fires for millennia.

Large deposits are found in Spain, Italy, and Peru. Pyrite can also be found in the U.S., particularly Illinois and Missouri.

Ancient Incans used pyrite to create mirrors, and archaeologists discovered more than 50 mirrors containing pyrite in an excavation in Arizona between the 1930s and 1960s. The mirrors were found buried with cremated human remains. Research showed the mirrors, found mostly broken, were made in Mexico.

These pyrite mirrors were returned to Mexico sometime after 2000. According to scientists, the Aztecs, who ruled during the 14th, 15th, and 16th centuries, spent hundreds of hours making one mirror. They believed the mirrors linked them with their ancestors.

During the Victorian Age in Europe, from about 1837 to 1901, pyrite was used to make jewelry. During World War II, pyrite was mined from the southern United States as a source of sulfur to make sulfuric acid.

PHYSICAL PROPERTIES

Pyrite forms in cubes and octahedrons, as well as flat disks called dollars.

MYTHS AND LEGENDS

Ancient Chinese believed pyrite would prevent crocodile attacks, and ancient medicine men were known to use it in their healing rituals.

Native American Indians considered pyrite to be a stone of magic and healing, and they also used the stones as mirrors, believing that the pyrite mirrors allowed them to peer into someone's soul.

USES IN HEALING

Healers say pyrite can reduce anxiety and depression and remove pollutants and negative energy from a room. One practitioner suggests putting pyrite on parts of the body that feel deprived of energy. Combining it with citrine, jade, or clear quartz is said to increase the effect of energy restoration.

Pyrite has also been touted to fight skin disease and infections caused by viruses or fungi, reduce the symptoms of blood disorders, and promote fertility.

Practitioners say wearing or carrying pyrite can shield the wearer from harm and danger. It can also be a source of creativity, not only in art, but also in math and science. Imagining a golden sphere surrounding you while carrying pyrite is said to instill new ideas and help you bring them to fruition.

Pyrite is touted to renew energy in someone who is fatigued due to long hours of work or study, and it is said to promote confidence and action in your desires.

Practitioners of the Chinese custom of feng shui use pyrite crystals to bring financial wealth. Feng shui relates to the belief that your surroundings can determine your level of harmony.

RHODONITE

Chakras: Root, Heart

NAMING, LOCATION, AND HISTORY

The name "rhodonite" comes from a Greek word meaning "rose," and refers to this stone's pink hue.

Rhodonite has been found in Sweden, Australia, India, Madagascar, Mexico, South Africa, Brazil, Canada, the U.S., and the Ural Mountains in Russia, where it was discovered in the late 1700s.

Local people called it the eagle stone because they said they saw eagles carrying small pieces of rhodonite to their nests. They began placing rhodonite into their infants' cradles as a protective stones, and travelers also began using it to keep them safe on their journeys. Rhodonite is the national stone of Russia. It is also the official gemstone of Massachusetts, where the crystal is mined.

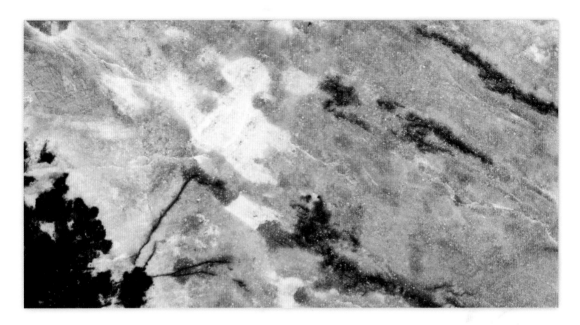

PHYSICAL PROPERIES

Rhodonite is pink and is sometimes found with a touch of brown. It also typically has some black streaks called inclusions. These inclusions occur due to the presence of manganese oxide. A polished piece of rhodonite can look like the inside of a watermelon.

USES IN HEALING

Rhodonite's sister stone, rhodocrosite, relates more to nurturing, while rhodonite, with its deeper color, is more empowering, according to crystal healers.

Rhodonite, like other pink stones, is known as a stone of love. It's said to bring love about by helping people use their talents to their fullest and discovering their true purpose.

Crystal practitioners also say rhodonite can help those who are trying to lose weight and detoxify. Some also say it can relieve symptoms of arthritis and emphysema.

ROSE QUARTZ

Chakras: Heart

NAMING, LOCATION, AND HISTORY

Rose quartz, a pink form of quartz, grows in small clusters or large formations and can be found in Brazil, Madagascar, India, and the United States. The state mineral of South Dakota, rose quartz has probably been in that region for at least one billion years, but it was first discovered in the state in the late 1800s. Some people collect rose quartz in the Buffalo Gap National Grasslands and the Black Hills National Forest.

Archeologists have discovered rose quartz beads in Iraq that were likely made in 7000 BC. Other rose quartz jewelry dating to about 800 BC has also been unearthed.

MYTHS AND LEGENDS

Ancient Romans and Greeks used rose quartz as a symbol of a covenant, and Greek and Roman myths mention rose quartz. For example, Cupid, the Greek god of love, gave the gift of love to humans with a rose quartz. Another myth says the color of rose quartz comes

from blood shed when the goddess Aphrodite was trying to save the god Adonis, who was being attacked by Ares. The myth says Aphrodite got pricked by a briar bush and that her blood stained white quartz pink. Zeus returned Adonis to Aphrodite every six months, making rose quartz the symbol of everlasting love.

Rose quartz has also been linked to the love of beauty and art in ancient times. Archeologists have discovered facial masks made of rose quartz in Egyptian tombs.

Ancient Egyptians believed rose quartz slowed the signs of aging.

Australian medicine men used rose quartz to create what was called sacred water, which was considered a cure-all. Native American cultures used rose quartz to stop anger and create love in its stead.

PHYSICAL PROPERTIES

Rose quartz ranges in color from pink to peach to rose, and some stones reflect star-like features in certain light.

USES IN HEALING

Crystal practitioners say rose quartz can be used to instill a sense of love, compassion, and kindness, and it can also help those with heart issues. They recommend placing a large piece of rose quartz in rooms of the house to maintain an energy of love.

It can be used to nurture by placing 12 of the stones evenly around the body, according to practitioners.

Healers suggest meditating with rose quartz to bring love and compassion into your life. Hold the rose quartz at your heart chakra and meditate, feeling all emotions that arise. Inhale love and exhale negativity while sensing a pink light moving through your body. You can also try repeating the mantra, "I will accept the light of love."

Rose quartz has also been said to relieve tension, stress, and anxiety. Though it's mostly related to the heart chakra, it can also help harmonize the crown, third eye, and throat chakras, according to practitioners.

RUBY

Chakras: Root

NAMING, LOCATION, AND HISTORY

Centuries ago, all red stones, including garnet, were thought to be rubies—but mineralogists later determined that ruby has its own specific chemical makeup.

Rubies come from Myanmar, Thailand, and several other countries. In Myanmar, miners collect rubies—as they have done for centuries—by chipping away at rocks and dirt to release the crystal. Many people prize rubies over even diamonds, and one story says a Chinese emperor wanted a large ruby so badly that he offered to pay for it with a city.

In Persia during the 1600s, an emperor had his throne decorated with gems, including more than 100 rubies.

Hindus consider ruby the king of precious stones. *Mani Mala*, a book about gems published in 1879 and considered a scholarly work about early culture, mentions an artifact entirely made of precious stones, including rubies, that was offered

to the gods. Hindus today leave offerings at temples, and those who leave rubies believe they will enter their next life as an emperor or a king. The book also links rubies to wealth, safety, and good fortune.

In the 20th century, man-made rubies were used to manufacture lasers. One of the world's most valuable cut rubies is heart-shaped and set into a necklace with diamonds. It's said to be worth about $14 million and contains a Burmese ruby.

MYTHS AND LEGENDS

Ancient Greeks believed rubies could boil water and melt wax. Other ancient peoples thought that by wearing rubies they could remain safe, happy, and peaceful. A Burmese legend says rubies inserted into the flesh would deem a person invincible. The most perfect ruby, according to the legend, was one that was the color of a pigeon's blood. Another legend says if a ruby is placed beneath a tablecloth, it will glow until it's removed.

Ruby has been thought to protect your home and land. In a book of travels from 1365, a story was told of a man who placed rubies on the four corners of his property to protect his home from violent storms and to increase his harvest.

PHYSICAL PROPERTIES

Rubies are the crystal form of the mineral corundum. Rubies are red due to the presence of chromium and

aluminum. Alternatively, sapphire, another crystal form of corundum, is blue because of the presence of chromium and aluminum.

Some rubies display bright, white star patterns atop the red stone. These are called star rubies.

USES IN HEALING

Ruby is said to improve self-esteem and confidence as well as add energy and vitality to the mind and the body. It's also considered an aphrodisiac, able to rekindle passion.

Crystal practitioners also believe that it has protective powers as well the ability to improve circulation and help with weight loss. Placing 12 quartz crystals around the body and then placing a ruby on the heart chakra is said to help people achieve their dreams.

RUTILATED QUARTZ

Chakras: All

NAMING, LOCATION, AND HISTORY

Rutilated quartz has been found in Brazil, Madagascar, and India, as well as other countries.

The term "rutilated" refers to the appearance of strands of hair in the crystal. Some other names for rutilated quartz include "Venus hair stone" and "Cupid's darts."

PHYSICAL PROPERIES

Rutilated quartz is typically found in a golden yellow color and is said to look like small bars of gold. It can also appear coppery-brown.

The strands in rutilated quartz can be thick, thin, parallel, and crossed, and are often gold or brown. Those with gold strands are called golden rutilated quartz.

USES IN HEALING

The stone is said to slow the aging process and help with healing, as well as decrease depression and take away loneliness. Practitioners also say it helps the hair grow and counteracts baldness. It's also used to help wounds heal more quickly.

This stone is said to help you let go of the past and rid your body of disease and negative energy. In addition, it's touted to foster forgiveness and help with life's transitions.

Pairing it with goldstone is said to help you get in touch with your instincts. Practitioners also suggest placing rutilated quartz on parts of your body with muscle strains to encourage healing. They also suggest placing it beneath a pillow at night if you're faced with a problem you can't solve. Upon awakening, healers say, you may have the solution. It also has been said to enhance your psychic abilities.

SAPPHIRE

Chakras: All

NAMING, LOCATION, AND HISTORY

Some of the largest deposits of sapphire are found in Thailand, Australia, Sri Lanka, and Madagascar. Australia was once the largest producer of this gem, until a large deposit was discovered in Madagascar. The stone is fairly easy to mine with hand tools by digging up some rock and then washing it on a screen and sorting through the remains.

Among the rarest sapphires is the Star of India, discovered in Sri Lanka three centuries ago. It has a star effect called asterism on both sides of the stone. It's housed at the Museum of Natural History in New York.

Another is the Black Star Sapphire of Queensland, purportedly discovered by a 12-year-old boy in the 1930s in Australia. Some news accounts say his father used it as a doorstop for several years before he realized its importance.

The Black Star Sapphire has rarely been seen in public, except when Cher wore it while starring in her television show, *The Sonny and Cher Comedy Hour*, in the late 1960s. It was also displayed at the Smithsonian Institute around the same time. Its worth has been estimated at $100 million.

Prince Charles of Wales gave Lady Diana Spencer an engagement ring featuring a blue sapphire set in white gold and surrounded by diamonds. Prince William kept the ring after his mother's death and gave it to his wife, Kate Middleton, when he proposed.

PHYSICAL PROPERTIES

Rubies and sapphires are varieties of corundum. The red variety is ruby and the blue is sapphire, although you can also find yellow, orange, pink, and white sapphires. Blue sapphires contain iron and titanium, while other colors have lower amounts of iron.

MYTHS AND LEGENDS

Rulers of ancient Persia believed the sky was painted blue by the reflection of sapphire stones.

USES IN HEALING

Blue sapphire is linked to the throat and third eye chakras. It's touted to ease headaches, eye problems, and vertigo as well as help foster a sense of security in one's knowledge.

Yellow sapphire, related to the solar plexus chakra, is thought of as a stone of prosperity and is said to help you feel the excitement of life and inspire a willingness to take risks. Its physical

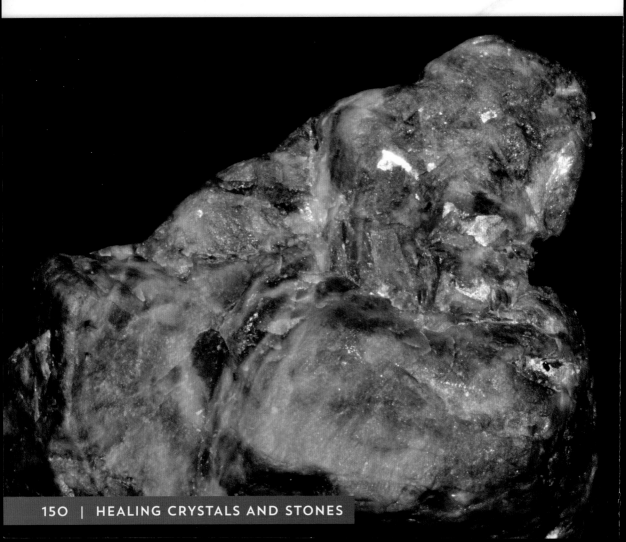

healing properties are said to include helping the digestive system and providing energy for those performing physically demanding work.

Orange sapphire is linked with the sacral and heart chakras. It's been touted to improve libido as well as enhance creativity in writers, singers, and other artists.

Pink sapphire works with the heart chakra and has been used by crystal healers on those with high blood sugar levels. Deposits of pink sapphire have recently been discovered in Africa.

White sapphire almost looks like a diamond. It works with the third eye and crown chakras to inspire courage.

SELENITE

Chakras: Third Eye, Crown

NAMING, LOCATION, AND HISTORY

Selenite was named after the Greek goddess of the moon, Selene, and it can be found in Greece, Mexico, Russia, Poland, England, France, and the United States.

MYTHS AND LEGENDS

Selenite is known as the moon stone and was used by ancient peoples to get rid of evil spirits. Ancient Mesopotamian cultures used selenite in rituals as well as to ward off evil. Hanging selenite on a fruit tree was also believed to increase the harvest.

PHYSICAL PROPERIES

Selenite, a type of gypsum, comes in gray, white, green, and brown hues. It can also be colorless. Selenite formed as salt water evaporated in lakes surrounded by land. Because of its softness, it can easily be scratched and requires care when handling. If left in water, it will slowly dissolve. Some specimens can even be bent with the hands.

USES IN HEALING

Selenite is considered a power stone that can rid the body of blockages that are keeping it from entering a higher state of awareness. It's often combined with other stones, such as tourmaline, for healing purposes.

Some practitioners recommend using selenite as a healing wand. Natural crystal wands of selenite are available for sale. One way to use one of these wands is by gluing crystals relating to the different chakras to it. Practitioners use selenite wands to help the body heal or restore balance. One healer recommends sitting in a chair with your bare feet on a selenite wand, then breathing in and out and feeling the earth's energy enter from your feet into your body.

SERPENTINE

Chakras: All

NAMING, LOCATION, AND HISTORY

Ancient peoples thought serpentine could cure the effects of snakebites, which is one way the stone may have gotten its name. Another possibility is that some specimens contain snakeskin-like patterns due to the crystal's magnetite content. Or perhaps it's simply because the color green resembles that of a serpent.

Serpentine is a source of asbestos and magnesium, and crystal healers say to only use serpentine in its rock form and not in its fibrous form. This statement is a reminder to take care with whatever product you use and to consult your physician before embarking on crystal healing.

Interestingly, serpentine is the state rock of California, and in 2010, lawmakers sought to change that, thinking it contained asbestos.

However, geologists said it's not hazardous unless turned into dust and released into the air.

A rare type of marble, which is mined in Ireland, contains crystals of serpentine and other minerals. The marble formed from sediment deposited in a shallow sea 600 million years ago in Ireland. The marble was used to build the Westminster Cathedral in London as well as the Galway Cathedral in Ireland. It was even shipped to the U.S. to be used in the Pennsylvania state building as well as College Hall at the University of Pennsylvania.

Serpentine is found in Great Britain, South Africa, Poland, and other countries including the United States. The Inuit people of the Arctic region created carved bowls out of serpentine to use with oil burning lamps. They also carved animal figures out of serpentine for trade. In fact, serpentine is said to be one of the stones used most often by these people. An artist community on Baffin Island north of Canada works with serpentine, which is abundant in this region.

MYTHS AND LEGENDS

Ancient peoples not only wore serpentine to protect themselves from snakebites, but also made sculptures out of it. The serpentine had to be in its natural state or it would lose its protective qualities. Ancient Egyptian books were said to be written on serpentine tablets.

PHYSICAL PROPERTIES

Overall the color of serpentine varies from yellow-green to medium green to forest green. It has been confused with jade, but it's a separate stone with its own mineral composition.

USES IN HEALING

Used in meditation, serpentine is said to help open the heart chakra as well as activate what's known as the *kundalini,* or fire energies. Crystal practitioners say placing serpentine on the crown chakra can clear all the other chakras and aid in relaxation. It's also been touted to remove toxins from the heart and lungs and help absorb nutrients such as magnesium and calcium.

Practitioners also suggest serpentine can help release the desire to do good work instead of focusing on personal gains, and that it can restore energy to the body.

SHATTUCKITE

Chakras: Heart, Throat, Third Eye

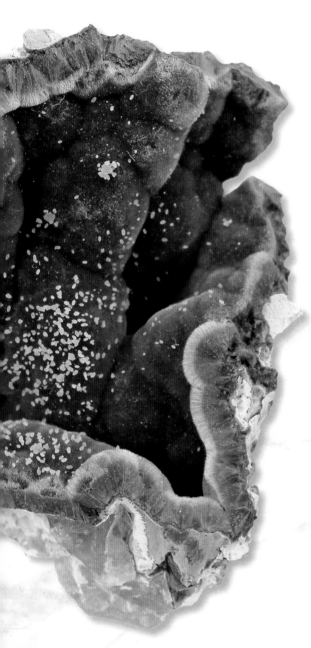

NAMING, LOCATION, AND HISTORY

Shattuckite was named after the Shattuck copper mine in Arizona where it was discovered in 1915. Other places where shattuckite has been discovered include Argentina and Austria. Some of the best specimens come from Namibia.

PHYSICAL PROPERIES

Ranging in color from medium to deep blue, shattuckite forms from the mineral malachite. During this formation, the crystal structure actually changes, although the shape does not. Specimens with both malachite and shattuckite in one stone can be found in Arizona.

USES IN HEALING

Placing shattuckite in the center of a room is said to create positive feelings and encourage optimism in those surrounding you. It also works with the throat chakra to help with communication skills, including those needed for public speaking and teaching, according to crystal healers. In addition, they say it helps humans understand the needs of their pets.

Crystal healers also suggest those recovering from major surgery and illnesses can speed the process by using shattuckite. They also say shattuckite, especially the dark blue form, is an important tool for those who believe in psychic powers and for those who want to deepen their meditation experiences.

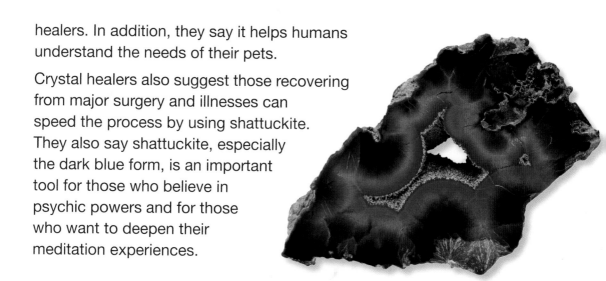

SMOKY QUARTZ

Chakras: Root

It's the national gem of Scotland, where Celtic peoples in 300 BC mined it and used it to create jewelry—including pins to wear on their kilts—and weapons. Chinese people in the 12th century used smoky quartz to make devices that would shield the eyes from the sun, akin to today's modern sunglasses.

NAMING, LOCATION, AND HISTORY

Smoky quartz is found in Australia, Brazil, Madagascar, and the United States. Crystals weighing more than 600 pounds have been found in Brazil, and some of the best examples of the crystal have been found in Colorado.

MYTHS AND LEGENDS

Ancient Druids believed smoky quartz reminded humans of the deep powers of gods and goddesses. Other cultures believed it could lead humans to the afterlife.

PHYSICAL PROPERTIES

Smoky quartz comes in colors from tan to dark brown. Clear quartz has also been artificially irradiated to produce smoky quartz.

USES IN HEALING

Practitioners today say smoky quartz can be kept in a purse or other container and carried while driving to protect against road rage and stress. Placing it around the home is said to protect the owner from theft and accidents, and it's been used to help with quitting smoking.

Others say it's best used to combat any ill effects of radiation. Sensitive people may wish to pair smoky quartz with clear quartz to calm the mind. Placing smoky quartz crystals around the body is said to create a sense of healing, especially in the hips, abdomen, kidney, and legs.

SODALITE

Chakras: Throat, Third Eye

NAMING, LOCATION, AND HISTORY

Sodalite deposits have been found in Brazil, Canada, India, and the United States. Its name likely comes from its high sodium content, or from a Latin word meaning "comrade." It has been said that wearing sodalite can make you feel as if you have a friend bringing you peace.

Huge deposits of sodalite were discovered in 1891 in Ontario, Canada, at the Princess Sodalite Mine. The mine was named after the Princess of Wales after she visited the World's Fair in New York in 1901 and was given sodalite as a gift. The stone was shipped to London, where she used it to decorate her home. The Princess Sodalite Mine Rock Shop in Bancroft, Ontario, offers adults and children the chance to search for the crystal and purchase specimens.

PHYSICAL PROPERTIES

Sodalite is typically blue-gray with white venation. It can also less frequently be found in green, yellow, and purple.

USES IN HEALING

According to practitioners, sodalite's healing properties include maintaining good blood pressure, improving digestion, assisting with weight loss, and reducing stress and anxiety. It's been called the poet's stone, and public speakers have been advised to use sodalite to help convey their message calmly and succinctly. Sodalite is also said to help bring out intuition and dispel self-doubt.

Practitioners also advise athletes to use sodalite to heighten endurance and writers to use it while seeking inspiration.

SUGILITE

Chakras: Third Eye, Crown

NAMING, LOCATION, AND HISTORY

Sugilite was named after Japanese geologist Ken-ichi Sugi, who discovered the stone in 1944. Most sugilite mined today comes from South Africa, but it can also be found in Canada, Italy, Australia, and India.

PHYSICAL PROPERTIES

Sugilite is a beautiful violet-colored stone that can sometimes be found in fuchsia, as well.

USES IN HEALING

Sugilite is known as a stone of protection and considered among one of the most useful crystals for those who tend to absorb whatever energy is around them. Crystalogists say carrying or wearing sugilite can protect you from negative vibes encountered within the home, at work, or in public.

This stone is also promoted as one of peace and harmony, aiding you in feeling connected and compassionate toward everyone. Placing it next to the bed at night is thought to encourage a peaceful sleep free of nightmares, while also instilling instructive and positive dreams.

Sugilite is used by crystal healers to relieve pain, especially headaches and joint pain. It is also used to fight viruses. Some crystalogists have suggested placing sugilite underneath the arms to remove toxins from the system.

Daily meditation with sugilite is said to help those with chronic pain. Meditating with amethyst and sugilite can increase a sense of protection and wisdom, according to crystal practitioners.

Sugilite has also been touted to help dispel negative thinking and improve self-confidence.

Some people believe sugilite can only heal those who have peace in their hearts and minds.

SUNSTONE

Chakras: Sacral, Solar Plexus

NAMING, LOCATION, AND HISTORY

Sunstone can be found in India, Norway, Russia, Madagascar, and the United States, among other countries.

The first collectors of sunstone in the U.S. were likely Native Americans, and Native American artifacts have been found containing sunstone. European settlers discovered sunstone in lava fields in a desert valley in Oregon, which has named sunstone its official state gemstone.

Ancient peoples used sunstone for trade as well as for navigational tools in the sea. Vikings were said to use sunstone to find the sun's position when it was hidden behind clouds or at night. Recent studies have shown sunstone very well could have been effectively used in that way.

PHYSICAL PROPERTIES

Sunstone crystals are orange-brown and embedded with hematite, which gives the stone an attractive sparkle.

MYTHS AND LEGENDS

Ancient Greeks believed sunstone brought abundance and warmth to those who owned or wore it, and it was considered a symbol of the sun god. Ancient Indians used sunstone to protect them from evil.

Native American legend says an arrow was shot into a warrior's skin and his blood colored the stones and gave them special powers.

USES IN HEALING

Sunstone is said to change anger into energy and inspire the wearer to offer service to others. It's also touted to raise the metabolism. Healers also suggest those who suffer from depression during cold, snowy winters hold a sunstone in their hand and look at its shininess to elevate the mood. It can also encourage the wearer to begin an exercise program, according to practitioners.

THULITE

Chakras: Sacral, Solar Plexus, Heart, Throat

NAMING, LOCATION, AND HISTORY

Thulite's name comes from the mythical island of Thule, which is said to be located at the farthest north end of the Earth. Ancient Greeks talked about Thule around 300 BC, and today historians think they were probably referring to Scandinavia.

Thulite was first discovered in a county in Norway with valleys and rock outcroppings in 1820. Sometimes called rosaline, the strawberry pink crystal is also found in Australia.

PHYSICAL PROPERTIES

Thulite is actually the pink variation of zoiste. Its rich pink color comes from a presence of manganese. It can often be found molten with white calcite.

USES IN HEALING

Thulite is said to calm those with a drama queen mentality and help keep a person's ego in check. Some crystal practitioners say thulite can be used by those who speak to large crowds and as a way to help build a rapport with others. It's also deemed an important crystal for strengthening the heart and the immune system.

One practitioner suggests placing thulite at the solar plexus chakra to calm the nerves or placing it at the sacral chakra to improve self-confidence. Placing four thulites around the body has been said to ease anxiety when facing a new situation.

Thulite is also touted as a way to make new friends and encourage creativity. Used in conjunction with obsidian and dumortierite, thulite is said to help break addictive behavior. Combining it with blue stones such as aquamarine and blue lace agate could also enhance communication skills, according to some healers.

TIBETAN BLACK QUARTZ

Chakras: All

NAMING, LOCATION, AND HISTORY

Tibetan black quartz is found in the Himalayan Mountains, which are considered a sacred location. In Tibet, monks gather the crystal by hand and place it into large bags. Crystal healers believe the Buddhist monks naturally bestowed sacred powers to these stones as they collected them.

Tibetan black quartz has also been discovered in Brazil, Russia, and Pakistan, among other countries.

PHYSICAL PROPERTIES

Tibetan black quartz is actually a clear stone, with some having thin layers of black. Tibetan black quartz is sometimes double terminated, meaning it has points on both ends. Double terminated crystals are often made into pendants.

Tibetan black quartz is also said to help meditators stay focused and remain relatively free from intruding thoughts. One Buddhist principal encourages you to remain unattached from emotions and desires. Tibetan black quartz is said to help achieve this state.

Placing this crystal beneath your pillow at night is said to protect you from having disturbing dreams.

USES IN HEALING

Some crystal practitioners consider Tibetan black quartz to be healing after surgery and helpful for boosting the immune system. In addition, some healers believe Tibetan black quartz soothes burns and provides relief to those suffering from vertigo.

Crystalogists suggest placing Tibetan black quartz in the four corners of the house and at the doors and windows in order to produce and maintain positive energy and promote protection.

TIGER'S EYE

Chakras: Root, Sacral, Solar Plexus

NAMING, LOCATION, AND HISTORY

Tiger's eye, also called tiger eye, can be found in India, Australia, Myanmar, and the United States. Some of the largest deposits are found in South Africa.

PHYSICAL PROPERTIES

Tiger's eye is a type of golden-brown quartz that sometimes contains blue streaks. When held a certain way, tiger's eye gives off a flash that some have related to what happens in the eye of tiger when it sees prey.

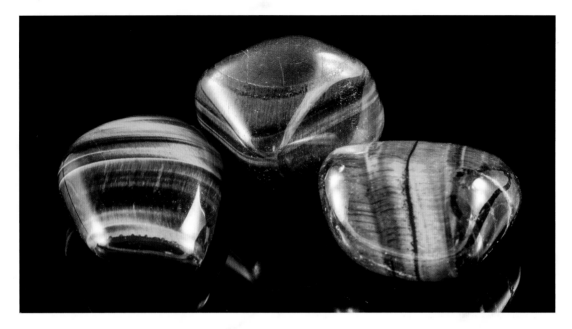

MYTHS AND LEGENDS

Ancient Romans carved tiger's eye into amulets for soldiers to wear to remain courageous and dedicated to the battle.

USES IN HEALING

Today, crystal practitioners believe tiger's eye brings energy to the body and helps heal eye and throat disorders. They also use it to balance emotions and encourage stability. It's additionally said to bring the wearer good luck and creativity.

Tiger's eye is touted to inspire harmony and patience and help the wearer recognize when the timing is right to complete a project, just as a tiger in the wild knows when to pounce on its prey.

A variety of tiger's eye, called falcon's eye, hawk's eye, or blue tiger's eye, has a dark blue-green sheen. It's considered a soothing stone, used to bring calm into the system as well as reduce the fear of flying and speaking. Some practitioners have used blue tiger's eye to help those who have difficulties with night vision.

TOPAZ

Chakras: Solar Plexus, Throat, Third Eye, Crown

NAMING, LOCATION, AND HISTORY

The name "topaz" comes from either the Sanskrit word "tapaz," which means, "fire," or a word meaning "to seek." In earlier times, any golden yellow stone might have been called topaz.

One of the largest producers of topaz is Brazil, where clear crystals weighing hundreds of pounds have been found. Blue topaz is found in Zimbabwe, and colorless to light blue varieties have been found in Texas. Other regions where topaz is found include Australia, Pakistan, Russia, Madagascar, and the United States. Utah named topaz its state gemstone in 1969.

A famous gem, the topaz of Aurangzeb, was amassed during the time of the Mogul Empire in Asia. Aurangzeb was one of its rulers, considered by historians today to have been oppressive and rich beyond measure.

Jean Baptiste Varnier, a 17th-century French gemologist was invited to view an impressive array of gems kept by the Mogul empire, and among them were topaz, diamonds, rubies, and pearls.

The American golden topaz, housed in the National Museum of Natural History in Washington, D.C., is considered one of the largest cut yellow topaz in the world. It was discovered in Brazil.

PHYSICAL PROPERTIES

Topaz comes in white, gold, blue, and pink varieties, depending on its chemical makeup.

MYTHS AND LEGENDS

Ancient Egyptians believed the god Ra created the gold topaz, and ancient Greeks believed topaz made them invisible. In the Middle Ages, people believed topaz could improve mental awareness and prevent insanity. People of India once believed wearing topaz above the heart chakra guaranteed intelligence and a long life.

USES IN HEALING

White or colorless topaz is said to work with the crown chakra and inspire confidence in telling the truth as well as the ability to tap into intuition and work with others on special projects. It's also touted to strengthen the hair and nails and help with skin issues.

Golden or yellow topaz is linked to the solar plexus chakra, and it is said to help those with urinary and kidney problems and help the wearer maintain boundaries. It's also said by practitioners to nurture new friendships. Some have claimed that placing golden topaz under the pillow can help you awaken with a positive, relaxed attitude. Placing topaz below the ribs has been said to relax and balance the body.

Blue topaz is linked to the throat and third eye chakras, offering a calming effect on the emotions and helping those with sore throats and migraines. Blue topaz is the state gem of Texas—it's rare to find a naturally occurring deep blue topaz in this state, so sometimes crystals are treated to produce that color. Blue topaz has been called the writer's stone and is said to help public speakers as well as those searching for their true paths in life.

TOURMALINATED QUARTZ

Chakras: All

NAMING, LOCATION, AND HISTORY

Also known as tourmaline quartz or black tourmalinated quartz, tourmalinated quartz is found in Brazil.

PHYSICAL PROPERTIES

Tourmalinated quartz is clear quartz with black tourmaline crystals that look like thin needles embedded within. With both these crystals present, tourmalinated quartz can help enliven and ground people, according to healers.

The crystal can be clear to somewhat translucent.

USES IN HEALING

Tourmalinated quartz is said to detoxify the body as well as help with digestive issues.

Placing tourmalinated quartz at the heart chakra is said to improve the immune system. Carrying it is touted to help with clear thinking, creating a type of screen around the body to protect it against negative energy.

It's known as an obstacle-removing

stone and has been touted to keep spaces free of electromagnetic radiation by placing it near the computer and television. Placing it in various rooms is said to keep toxins out of the environment.

Some practitioners have suggested sitting quietly, placing the stone on a cloth in front of the body and next to a candle, and then staring at the stone. They suggest letting random thoughts come to the mind and writing them down to review later. Tourmalinated quartz needs to be cleaned daily to rid it of any negative energy it may have absorbed during use. Cleaning practices include placing it in soil next to a healthy plant for several hours or leaving it in the moonlight for an evening.

TOURMALINE

Chakras: Root, Solar Plexus, Heart, Throat

NAMING, LOCATION, AND HISTORY

Tourmaline is found throughout the world. Tourmaline stones of different colors and mineral contents are found in different countries. This stone has been used since ancient times.

Russians in the 17th century erroneously called some tourmalines rubies. In Brazil, green tourmaline is called Brazilian emerald.

Tzu His, an empress of China from 1860 until 1908, was particularly enamored with tourmaline, especially the pink variety. She and her imperial court wore jacket buttons made of pink tourmaline, and many carvings were sculpted with the stone.

PHYSICAL PROPERTIES

Tourmaline can be found in different colors such as red, brown, black, green, and lilac. One type is called watermelon tourmaline and looks like the inside of the fruit.

When warming the crystal, one end becomes positively charged, and the other negatively charged.

USES IN HEALING

Tourmaline is considered one of the best stones for physical and spiritual healing, and it's been used since ancient times in many different ways.

Black tourmaline relates to the root chakra and is considered a protective stone, helping keep negative energies away as well as reducing stress, worry, and obsessive disorders. It's also been used to counteract the effects of radiation and excessive noise and to help those with lung diseases.

Pink tourmaline works with the heart chakra and varies from a pale pink to dark red. It's found in California as well as Africa and Afghanistan. It's considered a positive crystal, helping people deal with life's difficulties and live in harmony with the Earth.

The reddest variety of tourmaline is called rubellite. This stone is said to help those grieving from the loss of someone they love or trying to break free of abusive relationships. Rubellite, like black tourmaline,

has also been touted to calm the nervous system and help those with depression and obsessive disorders.

Green tourmaline can be light to dark, almost black, and is found in Brazil, Pakistan, Africa, and the United States. Compared with other tourmalines, it's considered a physical healer rather than an emotional one. It's said to be particularly effective when dealing with issues of the heart and is linked to the heart chakra.

Blue tourmaline, which relates to the throat and third eye chakras, and is said to help those with headaches and those trying to understand and resolve emotional traumas.

Brown tourmaline is linked to the root and heart chakras. It's also called dravite, likely named after Dravograd, a region in Slovenia where it was discovered. It can help with emotional healing, according to some practitioners, and also cleanse the lymphatic system. It can help heal digestive disorders such as irritable bowel syndrome, as well.

TURQUOISE

Chakras: Throat

NAMING, LOCATION, AND HISTORY

Turquoise, with its lustrous blue hue, has been mined since ancient times. The name "turquoise" likely came from a French word meaning "Turkish," relating to the trade in which the crystal was brought from Turkey to Europe.

Egyptians mined turquoise as far back as 3000 BC, when it was used to make amulets and jewelry. Cleopatra and King Tut were both enamored with the stone and were known to wear jewelry made from it. The death mask of King Tut is emblazoned with turquoise and other stones. Ancient Aztecs also made human skull masks inlaid with turquoise. Beads made of turquoise in 5000 BC have been discovered in Iraq.

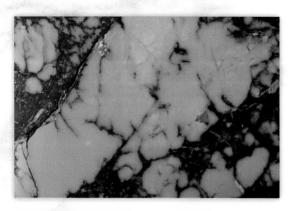

In 1810, Napoleon gave his betrothed Marie-Louise of Austria a tiara made of emeralds, which were later replaced with turquoise. The tiara is now housed at the Smithsonian Institute. The museum houses many Native American artifacts, including jewelry that contains turquoise.

Today, Native Americans continue to create pottery, art, and jewelry with turquoise. It is notably used to make fetish carvings of animals. People purchase turquoise necklaces to wear and carvings to keep in their home as protection.

Turquoise is found in Afghanistan, Australia, Iran, and other countries, including the U.S., especially Nevada and Arizona. Turquoise found in the U.S. typically has brown veining.

MYTHS AND LEGENDS

Native Americans threw turquoise into a river while praying to a god for rain. One legend says Native Americans created turquoise from the tears they shed when the rains came. Those tears mixed with the rain and fell into the earth to create turquoise. Turquoise represents life to Native Americans, and medicine men use the stone for healing.

The Navajos believed turquoise brought good luck, and they hung it in baskets to keep evil away. It was also given as gifts to promote friendship.

Many ancient cultures deemed turquoise a holy stone, and some thought if it changed colors while being worn, it foreshadowed danger. A Persian legend says when reflected in the new moon, turquoise can bring good luck. European men during the Renaissance wore turquoise rings to protect them from getting harmed during a duel.

PHYSICAL PROPERTIES

Varieties of turquoise found in the U.S. have different names, for example, Pilot Mountain turquoise from Nevada is blue to green and contains dark colorations. Royston turquoise from Nevada is olive green with dark hues.

USES IN HEALING

Crystal practitioners use turquoise to help with self-acceptance and to strengthen the heart, throat, and lungs. One healer suggests lying down with one turquoise at the head, one by the feet, one by each hand, and one by each shoulder. This can help you face obstacles and get rid of pain.

It's also said to help those suffering from depression, panic attacks, and physical exhaustion, as well as those with stomach issues. Wearing a turquoise necklace has been touted to prevent bronchial infections.

Those who meditate use turquoise for a deeper meditation session and to promote peace of mind.

TREE AGATE

Chakras: All

NAMING, LOCATION, AND HISTORY

Also known as dendritic agate, tree agate is found in India, Brazil and the U.S., particularly in Oregon.

MYTHS AND LEGENDS

Ancient Greeks buried tree agate in the soil after planting seeds to ensure a good crop, which could be where the name "stone of plenitude" originated. Ancient Greeks also associated the tree agate with the spirits of the woodlands. Tree agate is known as the agate of the goddesses, and it honors Roman and Greek goddesses of plants, the wind, the moon and other entities related to the Earth.

PHYSICAL PROPERTIES

Tree agate comes in white to gray to light purple and contains manganese or iron "dendrites" that look like trees or ferns embedded in the stone.

USES IN HEALING

Those who are seeking inner answers through therapy, meditation, and other programs

might find using tree agate to help, according to healers. Tree agate has also been purported to help relieve back pain when it's related to stress and tense emotions.

It's been touted to boost the nervous system and lower blood pressure. Placing a group of tree agates in a bowl in an air-conditioned home or workspace is said to fill the room with fresh air.

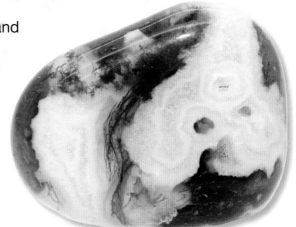

Various shades of tree agate simulate different chakras. Gray shades of this agate work with the root chakra, while a green tree agate is said to be linked to the heart chakra.

UNAKITE JASPER

Chakras: All

NAMING, LOCATION, AND HISTORY

Unakite jasper, or simply unakite, is a combination of feldspar, epidote, and sometimes quartz, and it forms in granite rock. Its name comes from the Unaka Mountains of Tennessee and North Carolina where it was discovered. It's also been found in South Africa, Brazil, and China, among other countries.

PHYSICAL PROPERTIES

The stone appears as a mosaic of green, pink, light orange, and sometimes red hues.

USES IN HEALING

Unakite is said to remove negative feelings and also speed healing from sorrow and grief. Those who feel overwhelmed with life may benefit from unakite, according to some crystal practitioners.

A stone of all chakras, but especially of the heart, unakite has been sprinkled with yarrow by some crystalogists and then placed in a bag to symbolize loving commitment. Unakite can also inspire love for all humans and nature, according to some, and instill a sense of togetherness or being one with the universe.

Unakite is additionally thought to improve healing of the lungs and heart.

Some healers use unakite to help those who want to quit smoking, improve eating habits, and reduce alcohol consumption.

Placing either a large piece of unakite or a bowl of small polished unakite stones in a room is said to instill calmness in that space, especially where televisions, computers, and similar devices are used.